A History of Groves

The grove, a grouping of trees, intentionally cultivated or found growing wild, has a long diverse history entwined with human settlement, rural practices and the culture and politics of cities. A grove can be a memorial, a place of learning, a site of poetic retreat and philosophy or political encampment, a public park or theatre, a place of hidden pleasures, a symbol of a vanished forest ecology, or a place of gods or other spirits. Yet groves are largely absent from our contemporary vocabulary and rarely included in today's landscape practice, whether urban or rural. Groves are both literal and metaphorical manifestations, ways of defining spaces and ecologies in our cultural life. Since they can add meaning to urban forms and ecologies and contribute meaningfully to the significance of place, critical examination is long overdue. The editors have taken care to ensure that the text is accessible to the general reader as well as specialists.

Jan Woudstra is Reader in Landscape History and Theory at the Department of Landscape at the University of Sheffield. His MA thesis at the University of York explored the wilderness in seventeenth- and early eighteenth-century gardens, and he has retained an interest in the topic, evidenced in various publications including 'The Early Eighteenth Century Wilderness at Stainborough', *New Arcadian Journal*, no.57/58 (2004–05), pp. 65–84. In 1997 he completed his PhD at the Department of Geography, University College London, with a thesis entitled 'Landscape for Living; Garden Theory and Design of the Modern Movement'.

Colin Roth is an interdisciplinary specialist in the psychology of aesthetics. His primary research field is in Danish culture through the 'long' nineteenth century, and his articles have appeared in the Royal Library Copenhagen's journals *Fund og Forskning* and *Carl Nielsen Studies*, more recently in a sequence for food professionals in *Chef* magazine. He has been a regular lecturer at the Tate and the Royal Academy of Arts. His editorial work began in the 1980s with the first *Journal* of the Derby Porcelain International Society, since when his wide disciplinary range has enabled him to work across the fields of musicology and ballet, art history, architecture and landscape. Dr Roth is currently co-director of the Centre for Nordic Studies at the University of Sheffield.

Routledge Research in Landscape and Environmental Design
Series editor:
Terry Clements
Associate Professor, Virginia Tech

Routledge Research in Landscape and Environmental Design is series of academic monographs for scholars working in these disciplines and the overlaps between them. Building on Routledge's history of academic rigour and cutting-edge research, the series contributes to the rapidly expanding literature in all areas of landscape and environmental design.

Regions and Designed Landscapes in Georgian England
Sarah Spooner

Immigrant Pastoral
Midwestern Landscapes and Mexican-American Neighborhoods
Susan L. Dieterlen

Landscape and Branding
The Promotion and Production of Place
Nicole Porter

Contemporary Urban Landscapes of the Middle East
Edited by Mohammad Gharipour

Melancholy and the Landscape
Locating Sadness, Memory and Reflection in the Landscape
Jacky Bowring

Cultural Landscapes of South Asia
Studies in Heritage Conservation and Management
Kapila D. Silva and Amita Sinha

A History of Groves
Edited by Jan Woudstra and Colin Roth

A History of Groves

Edited by Jan Woudstra and
Colin Roth

 Routledge
Taylor & Francis Group

LONDON AND NEW YORK

First published 2018
by Routledge
2 Park Square, Milton Park, Abingdon, Oxon OX14 4RN

and by Routledge
605 Third Avenue, New York, NY 10017

First issued in paperback 2022

Routledge is an imprint of the Taylor & Francis Group, an informa business

Publisher's Note
The publisher has gone to great lengths to ensure the quality of this reprint but points out that some imperfections in the original copies may be apparent.

British Library Cataloguing-in-Publication Data
A catalogue record for this book is available from the British Library

Library of Congress Cataloging-in-Publication Data
A catalog record for this book has been requested

ISBN 13: 978-1-03-240209-3 (pbk)
ISBN 13: 978-1-138-67480-6 (hbk)
ISBN 13: 978-1-315-56106-6 (ebk)

DOI: 10.4324/9781315561066

Typeset in Sabon
by Apex CoVantage, LLC

For our friends and families
and in memory of lost friends

Contents

Contributors

Camilla Allen trained as an illustrator and worked in publishing prior to completing her MA in landscape architecture at the University of Sheffield. This included a thesis on Richard St. Barbe Baker's work to combat desertification in Africa, which she is currently extending into PhD research, also at Sheffield.

Maureen Carroll is Professor of Roman Archaeology at the University of Sheffield, with particular specialisms in the archaeology of ancient Greek and Roman gardens, Roman death, burial, and commemoration, and Roman childhood. She taught at Cologne University and was senior field archaeologist at the Römisch-Germanisches Museum before coming to Sheffield in 1998. She has excavated Greek and Roman sites extensively in Italy, Cyprus, Tunisia, Germany, and Britain. Her important publications include *Der Garten von der Antike bis zum Mittelalter* and *Earthly Paradises: ancient gardens in history and archaeology*.

Yoshifumi Demura is Associate Professor in the Department of Civil Engineering at Gifu University, Japan. His research interests include the modern history of city development in Japan, civil engineering, and landscape.

Brent Elliott is a historian (formerly Librarian) of the Royal Horticultural Society. Author of *Victorian Gardens*, *The Country House Garden*, *The Royal Horticultural Society: a History 1804–2004*, and others. Formerly editor of *Garden History*; currently editor of *Occasional Papers from the RHS Lindley Library*.

Lei Gao is a researcher at the NMBU (Norwegian University of Life Sciences) in Ås, Norway. She is a former graduate of the University of Sheffield, where in 2010 she completed a PhD thesis entitled 'Breaking and Repairing: Conflicting Values in the Historic Gardens of China'.

Gert Groening was Professor of Garden Culture and Open Space Development at the Berlin University of the Arts from 1985 to 2009. He is a Professor Emeritus and continues to be research active at the Institute for History and Theory of Design, Berlin University of the Arts. He has held fellowships from the Beatrix Farrand Fund at the University of California in Berkeley; the Studies of Landscape Architecture at Dumbarton Oaks, Harvard University; the Japanese Society for the Promotion of Science; and Green College, University of British Columbia, Vancouver, Canada and has written various papers about heroes' groves.

David Jacques is a former Visiting Professor to De Montfort University and was Programme Director for the graduate courses in Landscape Conservation and Change at the Architectural Association in London. He is a conservationist and author on garden history; his latest book on *The Gardens of Court and Country* is a study of the formal landscapes of 1630–1730.

Dominique Juhé-Beaulaton is a researcher in history at the National Museum of Natural History (Paris–France). Her main research interest is landscape history in West Africa, with a focus on sacred groves as sources for local history and landscape construction in a long time depth. She has conducted a collective project on sacred groves and biodiversity conservation the results of which were published as Dominique Juhé-Beaulaton (ed.), *Forêts sacrées et sanctuaires boisés. Des créations culturelles et biologiques*. This project was financed by the French Institute of Biodiversity. She has also co-edited books about natural and cultural heritage and history of alimentation, and is currently studying natural history museum collections as sources for environmental history.

Hae-Joon Jung is an Assistant Professor in Landscape Planning at the Department of Landscape of Keimyung University, South Korea. He completed his PhD entitled 'Landscape as Heritage: Towards a Conservation Framework for Scenic Sites in Korea', at the University of Sheffield in 2015.

Colin Roth is an interdisciplinary specialist in the psychology of aesthetics. His primary research field is in Danish culture through the 'long' nineteenth century, and his articles have appeared in the Royal Library Copenhagen's journals *Fund og Forskning* and *Carl Nielsen Studies*, more recently in a sequence for food professionals in *Chef* magazine. He has been a regular lecturer at the Tate and the Royal Academy of Arts. His editorial work began in the 1980s with the first *Journal* of the Derby Porcelain International Society, since when his wide disciplinary range has enabled him to work across the fields of musicology and ballet, art history, architecture and landscape. Dr Roth is currently co-director of the Centre for Nordic Studies at the University of Sheffield.

Matthieu Salpeteur is a Marie Skłodowska-Curie fellow at the Centre d'Ecologie Fonctionnelle et Evolutive (CNRS, Montpellier) and Research Associate at ICTA-UAB (Barcelona, Spain). He has a PhD in anthropology with his research focusing on human-environment interactions. This is studied in different ways, through the dynamics of local ecological knowledge systems and contemporary evolution of mobility patterns among nomadic pastoralists (India); the historical ecology of sacred groves (Cameroon); and model simulation of complex social-ecological systems (India, central Asia). He uses methods ranging from comprehensive ethnographic surveys to quantitative tools (for example ecological knowledge indices and social network analysis). His publications include 'Espaces politiques, espaces rituels: les bois sacrés de l'Ouest-Cameroun', *Autrepart*, no. 55 (2010), pp. 19–38.

Marc Treib is Professor of Architecture Emeritus at the University of California, Berkeley and a noted landscape and architectural historian and critic who has published widely on modern and historical subjects in the United States, Japan, and Scandinavia. Recent books include *Settings and Stray Paths: Writings on Landscapes and Gardens*; *Representing Landscape Architecture*; *Spatial Recall: Memory in Architecture and Landscape*; and *Meaning in Landscape Architecture & Gardens*.

Tom Williamson is Professor of Landscape History at the University of East Anglia, where he leads the Landscape Group within the School of History. He has written extensively on landscape archaeology, agricultural history, environmental history, and the history of landscape design. His recent publications include *An Environmental History of Wildlife in England, 1550–1950*.

Jan Woudstra is Reader in Landscape History and Theory at the Department of Landscape at the University of Sheffield. His MA thesis at the University of York explored the wilderness in seventeenth- and early eighteenth-century gardens, and he has retained an interest in the topic, evidenced in various publications including 'The Early Eighteenth Century Wilderness at Stainborough', *New Arcadian Journal*, no.57/58 (2004–05), pp. 65–84. In 1997 he completed his PhD at the Department of Geography, University College London, with a thesis entitled 'Landscape for Living; Garden Theory and Design of the Modern Movement'.

Acknowledgements

The material for this book came first from a project and conference entitled 'Groves Lost, Found and Made', the work of a research cluster in the Department of Landscape of the University of Sheffield in 2013–14. Their aim was to explore the distinctive role and cultivation of groves in the past, and the making of new urban places through experimental art, design and management, re-working traditional practices. The histories of groves were studies from multidimensional and multicultural perspectives, using contrasting examples to explore why groves were created, how they were designed and planted, and how they were used. The year-long project culminated in a two-day conference, which together with the various seminars generated numerous interesting lectures and contributed to a debate. While not all contributions were directly translatable into written form, the debates which followed generated further offers of papers. The final selection for this book was dictated by the desire to provide a broad multicultural perspective and time depth.

Besides those here acknowledged as authors, other contributors to the debates and discussions included: Alison Hardie, then at the University of Leeds; James Bartos, now chairman of the Gardens Trust; Sylvie Nail from the University of Nantes; Patricia Debie from Renswoude, the Netherlands; Ian Rotherham from Sheffield Hallam University; Michael Klemperer, then at Wentworth Castle; Ana Duarte Rodrigues from the University of Coimbra, Portugal; Della Hook from the University of Birmingham; and the late Peter Blundell Jones of the University of Sheffield. We are grateful for their contributions. The colleagues in Sheffield who joined in the initial idea of research on groves and then contributed to this in various forms included Catherine Dee, Kamni Gill, Nicola Dempsey, Sally O'Halloran and Laurence Pattacini. The discussions were highly stimulated by a visit to the wildernesses and groves of Bramham Park, Wetherby on the final day of the conference, under the able guidance of Nick Lane Fox, the owner. Peter Goodchild, formerly of the University of York, was also there; in the mid-1980s he had encouraged me in my initial research on the topic of wildernesses in gardens.

We have enjoyed preparing this book together, sharing our complementary skills and experience, and hope this collection of essays reflects the breadth of interest, enthusiasm and curiosity of its authors.

Jan Woudstra and Colin Roth
Sheffield, February 2017

Introduction

Jan Woudstra

When the artist and poet Ian Hamilton Finlay (1925–2006) created his garden Little Sparta in the Scottish Borders, the primary structure evolved as a series of groves. This unusual approach was led by his interest in revolutionary politics, war and classical history, which provided a rich basis for interrogation and reinvention of what a grove was. It countered the general practice of landscape design and horticulture in which groves, as groupings of trees, intentionally cultivated or found growing in the wild, had rather lost interest in the oldest of man-made landscape features. They were seen as outdated, associated with ritualistic practices that were often derided as superstitious. Their long diverse history was entwined with human settlement, rural practices and the culture and politics of cities, and the grove had become overlooked.

Before Finlay the twentieth century had seen various revivals of the notion of groves, but these remained small-scale. The most popular types were memorial groves to commemorate an individual or more generally those lost at war. These had a particular international character, with many examples reported around the globe. In Mantua, Italy a grove was dedicated in the memory of Virgil in 1910.[1] In 1921 the landscape designer Madeline Agar planted a memorial grove at Wimbledon Common, London to commemorate World War I. It was designed as five of rings of forest trees crossed by two avenues of English oaks (*Quercus robur*), with a stone cross in the centre.[2] These trees were clearly an attempt to convey a symbolic meaning, and this tree had been used for millennia in north-western Europe for the same purpose. They were adopted in Germany to commemorate World War I in the so-called 'groves of honour', and a proposal for a national war memorial in Berka was referred to as rooted in 'pagan and poetical' ideals.[3] Gert Groening in this volume refers to them with a more direct translation as 'heroes' groves' and reveals how they were associated with nationalistic objectives.

More generally tree planting has been presented as a means of restoring land fertility, 'to plant hope and grow happiness', as it was described in the 1954 translation of *L'homme qui plantait des arbres* (The Man Who Planted Trees) by Jean Giono, which presented an idealized picture of one man's accomplishments in sowing a forest and revitalizing a whole land-scape.[4] In large-scale tree planting programmes groves were sometimes seen as a way to accomplish whole schemes, as for example in Palestine while under British jurisdiction, where in 1931 King George V planted the first tree in a forest on a hill near Nazareth. Not long after, groves had been established from donations from various countries, named 'in honour of some personage or event'.[5] Here the word grove was used to identify distinct donations and represented ownership as a way of transforming the ecology of the area, in the widest sense of that word. In Ireland official schemes that promoted reforestation encouraged 'all colleges and religious bodies to plant memorial groves to their own members'. The groves were intended as a means of underpinning the economy while generating a sense of national pride. This set a tradition which continues till today, largely as community driven projects.

In England, where there was not a centrally directed tradition for memorials of this kind, the most noteworthy twentieth-century groves were the 'coronation groves' in Windsor Great Park of 1937 and 1953. These were planted with oaks to represent the various territories of the empire, so 60 trees in the first and 62 in 1953, 'placed in a position corresponding with the territory's compass point in relation to the British Isles.' With the respective monarchs planting the first trees and others planted by representatives of the various territories, these were highly symbolic of their position and that of the empire. While this grove could be appreciated in plan, it is unclear how it might be appreciated as a distinctive feature, in contrast to, for example, the 'meditation grove' on top of a prominent hill at the entrance to Stockholm's Woodland Cemetery. Planted with characteristically shaped wych elm trees (*Ulmus glabra*) this is visually distinctive from its woodland surroundings and also provides the seclusion that invites contemplation. It is justifiably lauded as a modernist icon and has received worldwide exposure (Figure 1).[6]

The impression thus far of groves in the twentieth century is that they were designed features. Yet in America the best-known examples were naturally grown stands of Californian giant redwood trees (*Sequoiadendron gigan-teum*) which were referred to as growing in groves, possibly because of the majestic appearance and scale of these primeval forests. These groves have become known under separate names, one of them, the Bohemian Grove, notorious because of its association with the power network of American politics exposed by William Domhoff in *Who Rules America?* (1966).[7] Redwood groves are explored further in this publication by Camilla Allen.

Finlay, who questioned officialdom, sacredness, secrecy and purpose, may have known these examples. They would have formed a cue against which he defined and re-defined the notion of grove, which he explored in many of his 80 or so outdoor projects.[8] He communicated his perceptions, ideas and feelings in words, supported by image and metaphor, sculpted in concrete or stone. The metaphors he employed included a number of incongruous terms placed in relation with each other, using concisely written texts

Figure 1 The Meditation Grove at the Woodland Cemetery in Stockholm, planted with characteristically shaped wych elm trees, has become an icon of land-scape modernism (Jan Woudstra).

in epigrammatic form, clever and amusing, and providing new meanings. A 1982 installation entitled *The Sacred Grove* at the Kröller Muller Sculpture Park, the Netherlands included column bases set against existing trees representing a Pantheon. These included Lycurgus, Corot, Robespierre, Michelet and Rousseau; a lawmaker, a republican and a political philosopher, interspersed with a painter and a historian. These encourage us to rethink the relationships between art, history writing and politics, and distinguish an area of woodland that has been turned into a distinct grove by the act of hero worship and monumentalization. He replicated it in his garden at Little Sparta (Figure 2); an evocative poem embeds an invocation

Figure 2 Ian Hamilton Finlay's garden at Little Sparta includes a pantheon of the same cultural heroes he included in a 1982 project at the Kröller Muller sculpture park in the Netherlands (Jan Woudstra).

to a grove: 'THE SHADY GROVE/ THE MURMERING STREAM/ THE SHADY STREAM/ THE MURMERING GROVE'. An aspen grove contains a tablet in the shape of a gravestone with the inscription 'In a Sweet Harmony/ and Agreement/ with it Self/grove'. Finlay's poetic and physical deliberations provide a rich palette for further exploration of the notion of the grove, revealing his own profound understanding of their richness, and of how meanings are carried with words.

Defining grove

The origins of the grove are much older than its English name. In its Latin incarnation as a *lucus*, the sacred grove predates Christianity. The Romans clearly distinguished it from other woodlands that were not intended for religious purposes, such as *nemus, silva* and *saltus*. Over time the notion of *lucus* or grove changed, and they can be found in various forms and with different names in different locations, rising and declining in popularity as a result of socio-political circumstances. In this process of naming, intent and purpose all vary and sometimes take on nationalistic characteristics. In Old English documents dating back to 889, they are named *graf* and *græaf* (for thicket). The modern *Oxford English Dictionary* definition, 'a small wood; a group of trees affording shade or forming avenues or walks, occurring naturally or planted for a special purpose', even with the addition that they were 'commonly planted by heathen peoples in honour of deities to serve as places of worship or for the reception of images' is only part of the story. The origins of the word, unlike other terms describing woody planting and forest, are suitably mysterious and retain special significance. The word grove is English; in France, the word used is *bosquet*, which came from the Italian *boschetto*, or *bosco*. The Dutch word *bos*, or the diminutive *bosje*, also originates from this, as does the word bush in English, where it has a different meaning, as shrub.[9] In contrast in Germany the word *Hain* was used, and in Sweden *lund*, both with distinctive roots. In order to complicate matters even further, in early modern England the word 'wilderness' came into vogue to distinguish sacred groves from those with a profane, recreational function, and there were similar developments in other countries.

One interesting development is the use of the word 'grove' to describe various types of plantations of produce like lemon, orange, olive, nut or walnut, coconut and cherry, which today might be referred to as orchards. This is also true of groves of evergreens such as cedar, cypress, pine and yew, or evergreen broadleaves like box and holm oak. Plantations of single species of distinctive trees, such as ash, aspen, beech, birch, elm, laburnum, lime, locust, oak, maple and willow are sometimes referred to as being in groves; this practice may have started with citrus fruit. 'Groves' were associated with the Hesperides, nymphs of the sunset in ancient Greek mythology, and olive groves were incorporated in Plato's Academy near Athens. Classical Latin words ending with – etum or – arium referred to a grove or orchard, whether their trees bore fruit or not, such as pinetum for pine grove; populetum for poplar grove; salicetum, willow grove; ulmarium, elm grove; castanetum, chestnut grove; ficetum, fig grove; pometum or arium, fruit orchard.[10]

The underlying perception of groves has identified them with the sacred, the religious and mythical, which has been fashionable at some times but undesirable at others. The need for groves seems to be something deeply rooted in the human psyche. They stand for the importance of ritual, whether sacred or profane. So a grove can be a memorial, a place of learning, a site of poetic retreat and philosophy, or a political encampment, a public park or a theatre, a place of hidden pleasures, a symbol of a vanished forest ecology or a place of gods or other spirits. The importance of groves as a signifier and source of identity can be seen in their adoption for fieldnames, names for villages, houses and cemeteries. The word has also been adopted as one of the small cluster of names we use to give an identity to streets.[11]

While groves are otherwise largely absent from our contemporary professional vocabulary and have rarely been included in today's landscape practice, urban or rural, there have been some recent attempts to draw attention to them, most notably in Singapore. The city has been promoting itself as a City in a Garden, improving health and well-being through greenspace, a concept promoted by David Attenborough as a model for future cities.[12] The Gardens by the Bay were one of the key projects in a 2006 competition, won by a consortium led by Grant Associates. The focus of the project was on a grove of 18 'Supertrees', not real trees, but 25–50-metre-tall artificial structures that serve practical purposes such as providing shade or acting as rainwater collectors and chimneys for ventilation of buildings. They were planted with a range of epiphytes that could be viewed from stairways, and were described as vertical gardens. Here the notion of grove seems to have been used metaphorically, perhaps intended to reflect on the arrangement and draw attention to the ecosystem services provided by real trees (Figure 3).

Possibly inspired by this, one of the show gardens entered in the biannual Singapore Garden Festival held in 2014 was entitled Sacred Grove. On a small site where no digging was allowed, this consisted of a roof supported by a series of iron columns, angled in different ways in order to provide an impression of the chaos of the rainforest. In contrast, an irregularly spaced grove of 37 young exotic Brazilian ferntrees (*Schizolobium parahyba*) was planted on top of the roof; upright and all of equal size, they stood over a ground cover of grass and grass-like species. According to designers Wilson

Figure 3 The Gardens by the Bay in Singapore contain a grove of 18 'Supertrees', not real trees, but 25–50-metre-tall artificial structures that serve practical purposes such as providing shade, acting as rainwater collectors and chimneys for the ventilation of buildings (Alexander Patience).

McWilliam Studio they 'were playing with the idea of place and the creation of something special – a sacred grove that the ancients might understand'.[13] Although this provided a stunning image, it is not clear whether it would have been legible for the festival's Far Eastern visitors.

A history of groves

These examples illustrate some of the variety and richness in the interpretation of groves, explored in this volume. The emerging themes include sacredness, secrecy, seclusion, power, imperialism, nationalism, national identity, ownership and economic and ecological transformation, a rich palette that provides a wealth of entry points. This book emphasizes that it is their association with ritual, both sacred and profane, both daily routine and occasional event, that has rooted them so firmly in our cultures. They add meaning to both existing environments and new ones, creating new ecologies and contributing to the significance of place. Despite their cultural and physical significance they have largely escaped critical examination. The purpose of the present volume is to overcome this lacuna, and for the first time provide a set of papers that cover various phases in the development of groves from ancient times to the present.

Maureen Carroll in 'The sacred places of the immortal ones: Ancient Greek and Roman sacred groves' reveals that many cultures of the ancient Mediterranean world acknowledged the close relationship between the divine world and the natural environment; their temples and sanctuaries often included planted precincts that were considered holy and inviolable. The ancient Greeks and Romans, like their neighbours, also maintained and worshipped in sacred groves. This chapter examines various types of evidence for Greek and Roman sacred groves, including texts and inscriptions, pictorial depictions in reliefs and painting, and the archaeologically examined remains of plantings in an attempt to understand the function, use, location and appearance of sacred groves in Classical antiquity. It explores how sacred groves became enclosed and defined spaces in which mortals worshipped and communicated with the divine, and how increasingly they became symbolic of a benign pastoral world of old in which gods and men interacted in tune with nature.

Tom Williamson in 'Seeing the wood for the trees: the long-term aesthetics of woodland in England' explores the meaning of woods, groves and other forms of ornamental woodland in seventeenth- and eighteenth-century gardens. This needs to be understood not only in the context of high culture, classical and biblical texts and the like, but also against the backdrop of the long-term development of woodland management in England, especially the association of woods with elite status and their spatial relationship to high-status residences. Woods and groves were part of a long aesthetic tradition, and their increasing popularity in the period after the Restoration was one manifestation of a wider upsurge in enthusiasm for forestry and woodland.

David Jacques in 'The sacredness of groves' explores the reasons for the thoughts and emotions engendered in most people upon entering a mature grove of trees, remarked upon through the ages. Seventeenth-century England interpreted them in metaphysical terms, involving unseen but powerful influences. Some said it was communication with angels, providing inspiration and promoting true religion. Most authors on the subject searched the ancient authors for confirmation of their favoured causes and effects. The phenomenon of groves, and the writings of John Evelyn and others, are explored in order to give some order to the diverse writings which dried up in the last third of the seventeenth century.

Jan Woudstra in 'The history and development of groves in English formal gardens' explores how when Antoine Dézallier d'Argenville published his *La Theorie et la Pratique du Jardinage* (1709) it became the main source of reference on the topic; it was translated into English and German and there were pirated editions in the Netherlands. This book clearly categorized types of layout and planting, providing a vocabulary for contemporary garden design, including parterres and groves, recognizing and describing six different types of groves. This treatise clearly had an influence on later gardens, but it has also since been recognized as summarizing earlier trends. This chapter investigates the effect of this treatise in England. By identifying types and trends in the design of groves before the date of publication and afterwards it investigates innovation in the design of these features.

Brent Elliott in 'Colourful groves: the origins of the woodland garden' explores the fact that the vocabulary of 'groves' largely disappeared from gardening literature in the nineteenth century; in the wake of John Claudius Loudon, emphasis shifted from the mass to the gardenesque placing of the specimen. But in the third quarter of the century, renewed attention to arranging masses of trees resulted from a movement that arose for the planning of colour contrasts in the wider landscape. The eventual result of this was the woodland garden of the early twentieth century.

Camilla Allen in 'Groves as metaphor for the fragmented redwood forests of California' seeks an understanding of groves in the context of North American forests, and specifically through the contrasting stories of *Sequoiadendron giganteum* and *Sequoia sempervirens* groves that were identified in California in the nineteenth century. Two species, leftovers from another epoch: the former, otherwise known as giant sequoias, a collection of tall, barrelled, craggy grizzlies, nestled in the High Sierra Mountains; and the latter, a mist-shrouded expanse of forest made up of dense and slim trunks, the coast redwood. From the mid-nineteenth century onwards, the *Sequoiadendron* were rapidly discovered, exploited, celebrated and preserved, whilst at the same time, and with much less fanfare, the coast redwood forest was reduced to a fraction of its dominion before efforts were eventually stepped up to ensure its survival.

Dominique Juhé-Beaulaton and Matthieu Salpeteur in 'Sacred groves in African contexts (Benin, Cameroon): insights from history and anthropology' explore how in Africa, the expressions 'sacred groves' or 'sacred forests' are commonly used by a wide range of actors, from inhabitants to conservation practitioners and scientists. Behind these apparently consensual terms, a large variety of situations and phenomena can be found which make any generalization about the characteristics or functions of these sites uneasy and uninsightful. This chapter focuses on case studies drawn from field experiences in Benin and Cameroon. A close examination of the historical trajectories and of the functions of these sites shows that multiple forms of vegetation cover, of religious practices and ritual prescriptions, and management practices are indeed intertwined in these specific places. In the last decades, some new socio-political dynamics grounded on discourses about tradition revival, biodiversity conservation and cultural heritage have added new layers of meaning and functions to these sites.

Lei Gao in 'Groves in Chinese gardens and beyond them' discovers that there is surprisingly rich information on groves scattered in historical texts, which can roughly be grouped in four types: groves in myths and legends, groves in ritual places, fengshui groves and groves in gardens. In Chinese, 'grove', 'wood' and 'forest' share the same word '*lin*'. Sometimes another, or a few further words are added to describe what type of *lin* it is. However, there is not a clear distinction between forest and grove, and the fundamental difference is in the perception: a forest is seen as a wildness with unknown features and mysterious forces and hardly accessible to human beings, while a grove always has a relationship with people – it is planted, used, appreciated or worshipped by people. These qualities can be distinguished in all four types of groves. Ritual and fengshui groves are planted after certain patterns that represent ideal nature or heaven, and are therefore believed to contain the mysterious force that the pattern has. Garden groves are appreciated for their natural beauty, productivity and symbolic meanings. The groves of myth and legend appear to be magical, or a threshold leading to a magical world. Fengshui groves, ritual groves and garden groves can often find their origins in myth and legendary narratives.

Hae-Joon Jung in 'Korean village groves' reveals how village groves, or *maeulsup*, have played important cultural and educational roles in promoting a sense of identity, attachment and unity within the community in traditional Korean villages. Rural villagers planted village groves when they founded a new community based on their own culture and traditional ecological knowledge, and cooperatively owned, managed and protected them. These man-made groves were usually created at the *Sugu* area, the mouth of the village's watershed, taking on a threshold function, so they have been regarded as an intermediate space between village and nature, sacred and secular. Playing a major part as the setting of the village, they have been perceived not only as a sacred place in which to pray for a village's prosperity, but also as resting places and for entertainment, like today's public parks. More than a thousand groves remain in South Korea today, but many of them have been degraded and even destroyed during recent decades of industrialization and urbanization. Yet village groves are an important part of the heritage and a distinctive type of cultural landscapes. In order to safeguard their cultural and ecological diversity, they should be systematically conserved and awareness of their value risen by engaging the wider public.

Yoshifumi Demura in 'The shrine groves of modern Japan' reveals that although all traditional Japanese Shinto shrines have dense 'groves', the significance of the shrine groves was only discovered or recognized in the struggle to fashion the modern Japanese nation. In this chapter, two parallel events are examined: the shrine merger movement and the construction of the Meiji Shrine, which both influenced perception of the shrine groves. The way in which the Meiji government attempted to establish a modern autonomy using the values and virtues of shrines is subjected to scrutiny. The values included the practical existence of the grove as a physical place connected or emotionally attached to villagers' lives, reverence for their ancestors, a common history and an environment as a basis of local autonomy. These ideas were venerated within a specific place: the eternal evergreen grove. In pursuit of modernity this was presented as superstition. By the time of the creation of a Meiji Shrine, two kinds of values were established: one involved the preservation of the shrine precinct as a divine place; the other emphasized the importance of it as a public park. Ultimately preservation of shrine groves was incorporated into city planning policies, in which the image of the eternal grove was diluted into a zone described as scenic or green.

Gert Groening in 'Nature mystification and the example of the "heroes' groves" in early twentieth-century Germany' explores how the heroes' groves concept of early twentieth-century Germany carries aspects of politics and *Weltanschauung*. The landscape architect Willy Lange connected the traditional symbol of an oak to a national-monarchic context. With his idea of an oak grove Lange related to holy groves of Germanic antiquity. He tried to legitimize a continuity of German history from Germanic prehistory to the world-political claims of Germany during World War I. Willy Pastor, a German cultural historian, supported heroes' groves on similar lines. He referred to a circular motif, which he considered characteristic of a religious testimony to an earlier Germanic empire. Pastor considered the inhabitants of this empire superior to other people, both culturally and racially. Johannes Speck, a teacher, transferred the heroes' groves concept to military purposes. For him heroes' groves were to enhance the emotional and physical military readiness of young men, acting as an instrument in Germany's quest for greater power. The earlier idea of heroes' groves as sites of commemoration of ancestors, which had given rise to the concept, was encompassed within this ideological intention. It appears that its outstanding success both with state institutions and citizens rested on the way it was staged politically. Lange declared heroes' groves as a kind of national holy place for church-related and state ceremonies in which heroes' groves functioned as a church, temple, altar or chapel. Speck hoped that heroes' groves would become centres and carriers of national and religious life, adding another dimension to their 'holiness'.

Marc Treib in 'Dan Kiley: groves, space and complexity' recognizes that our first associations with the word grove might be more poetic than factual, involving clusters of trees in naturalistic clumps, perhaps enlivened by a spring and even a nymph. More pragmatically, we might think of trees set in regular intervals grown for productive purposes. These two orders, the natural and the geometric, establish the limits for contriving a grove; both have appeared regularly throughout at least two millennia of landscape architecture. Rather than examining trees in history, however, this paper discusses groves designed in the twentieth century by Daniel Urban Kiley, focusing on the *spaces* he achieved by planting in a geometric order. Military service in Europe at the end of the Second World War provided Kiley with the opportunity to discover garden cultures new to him, in particular the landscapes of André le Nôtre. In them he found the beauty of order and its ability to structure, and humanize, space and modulate scale. To Kiley the geometric bosk or allée was as natural as the irregular clump, believing that it is not man *and* nature, but that 'man *is* nature, like trees'. Four Kiley landscapes will illustrate his evolving use of the grove and how the spaces that resulted from a regular ordering became more complex, sophisticated and engaging over time.

Notes

1 An., 'Memorials to great men', *The Times,* 1 April 1910, p. 11.
2 An., 'Rings of forest trees: Wimbledon Common War Memorial', *The Times*, 25 April 1921, p. 5.
3 An., 'Germany's War Memorial; perpetual fire in the Grove of Honour', *The Observer*, 30 May 1926, p. 12; 'The most German spot: Berka for the National War Memorial', *The Observer*, 18 July 1926, p. 10.
4 The initial translation of the book was *The Man Who Planted Hope and Grew Happiness* (London: Conde Nast, 1954).
5 An., 'Memorial trees: Jews' tribute to King George V', *Manchester Guardian*, 27 September 1938, p. 11; An., 'Ignored', *Irish Independent*, 23 August 1979, p. 10.
6 It became iconic through a photograph first reproduced in Great Britain in Peter Shepheard, *Modern Gardens* (London: Architectural Press, 1953) which opens with it, on p. 25; see also Caroline Constant, *The Woodland Cemetery: Toward a Spiritual Landscape* (Stockholm: Byggförlaget, 1994), p. 10; now a World Heritage Site.
7 Harold Jackson, 'Playtime and politics down in Bohemian Grove', *The Guardian*, 20 July 1982, p. 6.
8 For Finlay see: Patrick Eyres, 'Garden, park and cityscape: A checklist of Ian Hamilton Finlay's permanent landscape installations', *New Arcadian Journal* 61/62 (2007), pp. 55–82; pers. com. Patrick Eyres, 10 December 2016.
9 'Bush' retains its English sense of 'thicket' when it describes a place of concealment where one might lie in ambush, notably in a colonial context.
10 David G. Pattison, 'Suffixed tree-nouns and grove-nouns in early old Spanish', *Neophilologus* 59 (1975), pp. 242–53; Walter E. Geiger, 'Fruit', 'Fruit Tree', and

'Grove' in Spanish: A study in derivational patterning', *Romance Philology* 20/2 (1966), pp. 176–86 (181).

11 See for example: 'Call it Grove, never Street', *The Guardian* 5 August 1971, p. 5.
12 David Attenborough in Planet Earth II, episode 6 'Cities', BBC, December 2016.
13 Pers.com. Andrew Wilson, 30 November 2016; www.wmstudio.co.uk/show-gardens/#/sacredgrovesgf2014/

1 The sacred places of the immortal ones

Ancient Greek and Roman sacred groves

Maureen Carroll

Introduction

Many cultures of the ancient Mediterranean world acknowledged the close relationship between the divine world and the natural environment, with trees and vegetation being interpreted as a sign of God-given life. Accordingly, the temples and sanctuaries of many gods in ancient Assyria, Egypt, and Israel, to name but a few locations, included planted precincts that were considered holy and inviolable.[1] The ancient Greeks and Romans, like their neighbours, also maintained and worshipped in sacred groves. An early expression of the belief that 'the sacred groves of the immortal ones' were the dwelling places of semi-divine beings can be found in the *Homeric Hymn* to Aphrodite; in the groves of the *Hymn* lived the nymphs who were born as trees.[2] The Greeks also had notions of paradise in a mythical grove of plenty and immortality.[3] The importance of the ancient sacred groves of the heroic past in the literature and thought of Roman Italy in the late first century BC also reflects a cultural and religious reverence for ancient trees and rustic landscapes.[4] Although secular groves of trees and orchards also existed in Greco-Roman antiquity, this paper focuses on those connected with cult sites, simply because they are more abundantly attested.

The types of evidence available for the study of Greek and Roman sacred groves are varied, as this brief introduction attempts to show. One category of material is the textual evidence in the form of written accounts and inscriptions. Descriptions of venerable sacred groves of olives, pines, cypresses, oaks, laurels, and fruit trees can be found in Pausanias' guide book of ancient Greece in the later second century AD, and he refers to them as some of 'the most memorable and interesting things' for Roman cultural tourists to visit.[5] And because of the text of a Roman inscription of c. 300 AD, we know that the ancient sacred olive grove associated with the temple of Athena on the acropolis of Lindos on Rhodes was renewed with plantings donated by the priest Aglochartos.[6]

There is also the pictorial evidence. A marble funerary relief of c. 100 BC from the eastern Mediterranean, now in the Getty Museum, for example, depicts a woman and four young girls in a wooded grove.[7] A single tree with five large leafy boughs is depicted, which is a common shorthand and space-saving reference to a whole grove in Greek sculpture and painting. That this grove is sacred is evident by the presence of a sacrificial altar and a pillar with a votive offering (a *lekythos*) from a mortal suppliant on top of it. A belted dress as another votive offering is suspended from the tree.

Finally, a valuable body of material for the study of sacred groves in the Greek and Roman worlds is the archaeological evidence. Excavations at the Greek temple of Zeus at Nemea, for example, revealed the planting pits for twenty-four trees dug into the rocky subsoil of the sanctuary's precinct in the fifth and fourth centuries BC.[8] Charred remains of cypress wood at the bottom of the pits actually confirmed Pausanias' description of the grove as one consisting of this species of tree.[9]

In the subsequent sections, these main strands of evidence for sacred groves in Classical antiquity will be explored in more depth in an attempt to understand their function, use, location, and appearance.

Defining a sacred grove: the written sources

It is important at the outset to establish what ancient terms were used to refer to sacred groves and what properties such groves were thought to possess. In addition, it will be essential to this discussion to establish what activities took place in the groves 'of the immortal ones' and what provisions existed for their protection.

In the Greek texts, ἄλσος (*alsos*) is normally the word used for 'sacred grove', although this also could apply to a holy place, with or without trees.[10] Κῆπος (*kepos*) is generally used for 'garden', whether it was sacred or not.[11] The distinction between 'grove' and 'garden' is not always clear-cut; a grove of trees could be called a κῆπος, especially if it had fruit trees. The Latin texts use either *nemus* or *lucus* for a grove. Although scholars do not always agree on the distinction between the two, a *lucus* generally appears to have been a grove that was created and inhabited by divine spirits and was left in a natural state, whereas *nemus* refers to a grove created or manipulated by man and furnished with sacred buildings and images.[12] *Hortus* refers to a garden, but it appears to be used only in reference to secular gardens.

As suggested by this brief analysis of Greek and Latin terms, sacred groves could be natural woodlands or entirely man-made plantings of trees. Both types of sacred groves are attested by literary descriptions and by archaeological enquiry, although, as Edlund points out, even so-called natural open-air cult sites were generally shaped by man, be it through their enclosure within a boundary wall or their embellishment with an altar or sacred images and so on.[13] A relevant inscription from Rome sheds light on the creation of a sacred grove. The city of Rome was laid out on seven hills, each of the separate hills originally distinct and occupied by population groups, the Montani (hill dwellers). The Montani of the Oppian Hill had an inscription carved in the first century BC which recorded that the mayor (*magister*) and priests (*flamines*) of this community had been responsible for enclosing their central shrine and planting trees in it.[14]

Throughout the Greek and Roman worlds, sacred groves were religious spaces, marked out of the landscape from non-sacred land, and they were places where mortals worshipped and communicated with the divine. The ancient Greek verb τέμνειν, to cut out or to mark off, is reflected in the word *temenos*, used for sanctuary space marked off by a wall or a boundary stone.[15] In Roman religious law, ownership of the land was transferred ritually through *consecratio* to the deity, thereby removing it from secular access.[16] In these spaces, not only the trees but also, on some occasions, the living inhabitants of the groves possessed sanctity. The sacred grove (*lucus*) of Lacinian Juno six miles from Croton in southern Italy, for example, was enclosed by dense woods and tall fir trees, and in the midst of the grove were pastures for cattle sacred to the goddess.[17]

Some inscriptions record measures taken to protect sacred groves. Particularly interesting in this regard are two Greek decrees from the island of Kos, one dating to the end of the fifth century, the other to the fourth century BC.[18] The texts reveal that the cypress trees in the sanctuary of Apollo Kyparissios and Asklepios were not to be cut down and it was prohibited to remove wood from the precinct; any infringement was punished with a stiff monetary fine of 1000 drachmas to the authorities and offenders were deemed guilty of impiety against the temple. These measures had, in the first instance, a practical aim, the preservation of the trees in a time of general deforestation in Greece and an increased demand for sources of timber, but they also had a religious aim in protecting the sanctity of the temple's property.[19] Also Roman epigraphic evidence allows us to recognise rules and principles for preserving sacred groves, and these have much stronger religious aims. According to inscriptions and literary references, the Romans considered violations of sacred groves religious offences.[20] A municipal decree from Spoletium, dating to the period after 241 BC, for example, prohibited the removal of anything belonging to the sacred grove and the cutting of wood in it, except on the day of an annual festival.[21] Any violation of the grove was punished by an expiatory animal sacrifice directly to Jupiter and the payment of a fine. Cutting wood in a sacred grove might even be punishable by death, as it was for Decimus Turullius who felled trees in a sacred grove on Kos to acquire timber for the fleet of Mark Antony in 32–31 BC; the victor over Mark Antony, Octavian, had the man executed.[22]

Although the destruction or mutilation of sacred wooded sites was unacceptable and punishable, there are recorded incidents of this kind of intentional damage during times of war, and there is no information on how any rules or laws governing the protection of sacred groves might have been applied in these cases. King Philip V of Macedon, for example, attacked Pergamon on the western coast of Asia Minor in 201 BC, ordering his army to cut down the trees in the sanctuary of Athena Nikephoros, the city's presiding deity and dynastic protectress of Pergamon's Attalid kings.[23] The Greek gymnasia and schools in the suburbs of Academy, Lykeion, and Kynosarges outside Athens were situated from the fifth century BC along the banks of the Kephissos, Eridanos, and Ilissos rivers in the midst of ancient sacred groves and shrines, some of which were 'natural', others man-made.[24] The plane, elm, poplar, and olive trees in the Academy district, in particular, were praised in many ancient sources.[25] These institutions were highly revered even much later by aristocratic Romans seeking tuition in rhetoric and philosophy there, but not all Romans viewed these sacred groves as inviolable. The Roman general Sulla, in his attack on Athens in 86 BC and in need of timber for his siege engines, felled the trees in the shady groves of the Academy and Lykeion districts, regardless of their antiquity and importance to the Athenians.[26] A sacred grove close to the Greek colonial city of Marseilles in southern Gaul was cut down by Julius Caesar after the inhabitants had sided with his rival, Lucius Domitius Ahenobarbus, in civil war. The poet Lucan seemed to excuse this particular act of sacrilege because in this grove, which could not be penetrated by sunlight and in which no animals or birds lived, the Gallic and 'barbarous rites' of human sacrifice were practised, a custom repellent to the Romans.[27]

Sacrifices took place and religious rituals and festivals pertinent to the cult were celebrated in sacred groves, and rules determined who should enter the groves or perform tasks and rituals in them.[28] Some of the rituals are so shrouded in mystery and obscurity, particularly those of very old cults, that we cannot truly comprehend their origin or meaning. A prime example of this is the sacred grove (*lucus*) of Diana on Lake Nemi in the Alban hills south-east of Rome where an old, archaic temple to her was located.[29] According to Roman mythology, the Trojan prince Aeneas, fleeing from the ruins of Troy, visited this sacred grove and plucked a branch from a tree in it, going on to fulfil the divine prophecy of being the founding hero of Latium and the Roman people. This act became a ritual central to the religious office of the priest known as the *Rex Nemorensis* (King of the Grove). The *Rex Nemorensis* could be challenged by any runaway slave who, if able to break off a limb from a tree in the sacred grove and defeat the incumbent priest in armed combat, would become the King of the Grove.[30] By the time Roman authors were writing about the ritual in the first century AD, however, there was certainly no longer any aspect of real physical combat involved in choosing the *Rex Nemorensis*, but the grove continued to be an important aspect of the cult, as it is represented in abbreviated form on Roman coins as late as the first century BC.[31] Furthermore, women seeking help in conceiving a child in the first centuries BC and AD worshipped here as suppliants of Diana, leaving garlands and votive tablets hanging from the trees in the grove.[32]

In Rome itself, the Arval Brethren, an ancient fraternity of twelve priests, conducted archaic fertility rites and worshipped in a sacred grove (*lucus*) of the goddess of fecundity, Dea Dia, on the road from Rome to Ostia. One of their tasks may have been an annual ritual circuit of the Roman fields, as part of the Ambarvalia festival in May. Something is known about the rituals conducted in the grove, largely from inscriptions and texts from the early first to the mid-third centuries AD.[33] According to these sources, the grove contained ilex and laurel trees, and only the priests entered it to celebrate the annual sacrifice or to prune damaged trees or remove or burn off dead ones. The *lucus* remained inaccessible to the public.

The sacred grove of ancient cypresses on a hill above the source of the river Clitumnus in Umbria was, on the other hand, very much accessible to visitors. According to Pliny the Younger at the end of the first century AD, it was the site of an 'ancient and venerable temple, in which is placed the river-god Clitumnus clothed in the usual robe of state', but also 'several little chapels are scattered round, dedicated to particular gods'.[34] Pliny also mentions the many coins tossed by the visiting faithful into the crystal clear water of the sacred spring. The site attracted some visitors of very high standing, including the emperor Caligula who 'on a sudden impulse' went 'to visit the river Clitumnus and its grove' around AD 39.[35]

A direct connection was made between types of trees in sacred precincts and the deities worshipped there. Pliny the Elder remarked on the antiquity of the custom of the veneration of trees in temple precincts in his native Italy, saying that 'different kinds of trees are kept perpetually dedicated to their own divinities, for instance the oak to Jupiter, the bay to Apollo, the olive to Minerva, the myrtle to Venus, the poplar to Hercules'.[36] Trees in sacred groves were often believed to have magical or prophetic properties, some of them relating to the varying political fortunes of the empire. A fig tree growing in the forum in Rome was a sacred reminder of the tree under which the she-wolf had suckled the abandoned infants Romulus and Remus, the mythical founders of the city.[37] Whenever the tree shrivelled up it was thought to be a portent, and it had to be replanted more than once by the priests. Also in Rome, the shrine of Quirinus was one of the most ancient temples in the city, and in front of it two sacred myrtles grew for a long time.[38] One of them, the patricians' myrtle, flourished as long as the Senate did; the other one, the plebeians' myrtle, later grew stronger, when the influence of the plebeians dominated over that of the Senate. Ovid also mentions a shady grove in conjunction with this sanctuary.[39] At Nocera in Campania in south-west Italy, the growth of an elm in the sacred grove of Juno was related directly to the fortunes of the Roman Empire. It had been cut back during the wars against the Germanic Cimbri who swarmed into Gaul and northern Italy in the late second century BC because its branches hung down onto the altar, but the elm recovered immediately and began to flower, and from that time on 'the power of Rome recovered after being ravaged by disasters'.[40]

Sacred groves also were places of refuge and asylum in which divine protection against injury and injustice was sought. The altar and grove of the Twelve Gods in the Athenian agora, for instance, functioned as a place of supplication when, in 519 BC, the Plataeans sought protection there from the Thebans, placing themselves under Athenian protection.[41] The site continued to be a place of refuge on various occasions in the fifth and fourth centuries BC.[42] The Roman writer Statius referred in the first century AD to this grove as a *nemus* of olive and laurel trees, but the term 'grove' in this case clearly applied to a very small number of trees, as excavations conducted there in the 1930s revealed only three or four tree pits.[43]

Groves and gardens connected with the gods of a city were viewed as a reflection of heavenly power and sanction of the city and its people. This is clear in the connection made by Diodorus Siculus in the first century BC between the fertile groves and gardens near Panara on Sicily and the divine majesty of the place.[44] Sacred groves and temples could be so closely associated with the city or landscape in which they were located that they came to be considered a symbolic equation of that place. The grove of the temple of Apollo at Daphne outside Antioch on the Orontes in Syria, for instance, is depicted summarily on an illustrated map of the Roman world in the fourth century AD as a landmark of Antioch.[45] The city of Antioch is depicted as an enthroned female figure, and she is surrounded by the trees of her famous grove.

But groves were not always necessarily the location only of religious or cultic activities. They occasionally had a more prosaic role to fulfil because the temples could own estates that generated income. Within the walls of Athens, in the south-east corner of the city, for example, was a *temenos* of Neleus and Basile planted with over two hundred olive trees providing revenue for the cult.[46] Important details of this planted *temenos* and the sanctuary (*hieron*) of Kodros, Neleus, and Basile are recorded in a decree of 418–417 BC. The *hieron*, nearby by not contiguous, was not leased for cultivation because it was sacred, whereas the *temenos* could be because it was under public jurisdiction and designated for private use.[47] Such temple properties could be located at a considerable distance from the sanctuary. Numerous inscriptions recording temple property of this kind in Classical and Hellenistic Greece survive, especially on the island of Delos where officials of the sanctuary of Apollo managed its estates on that island and on the neighbouring islands of Rheneia and Mykonos.[48] Temple land could be a rental commodity. The estate of Zeus Temenites on the island of Amorgos, for instance, was leased contractually to a tenant who tended to its orchard of fig trees; he also was responsible for erecting and maintaining a boundary wall around this plantation.[49] The tradition of maintaining estates worked by tenants for the profit of the temple continued in the Roman period. A Greek inscription of the fourth century AD from Herakleia in Sicily preserves a lengthy list of temple lands, including olive groves, vineyards, and woodlands.[50] The tenants were responsible for planting vines and trees and caring for them, as well as replacing old and unproductive vines or trees with new ones.

Images of sacred groves

Sacred groves filled with sanctuary furniture, gods, and worshippers engaged in cult activities are sometimes shown in Greek and Roman painting and sculpture. Some of the earliest Greek depictions in painting are those on red-figure vases of the fifth century BC produced in Athens. On one of these vessels, a wine mixing vessel (*krater*) of c. 440–420 BC in Agrigento, the god Apollo is seated in his sacred *temenos*, indicated by a tripod on a column, an altar, and a single, leafless tree.[51] Worshippers lead in a goat to be sacrificed on the altar. Roughly contemporaneous is a red-figure bell *krater* again showing Apollo in front of what is recognisably a laurel tree beneath which is a blood-spattered altar.[52] A youth roasts sacrificial meat for himself and the other worshippers on a spit over the flames on the altar, whilst another pours wine from a jug onto the altar.[53] Elsewhere, on a fragmentary bell *krater* of the third quarter of the fifth century BC in the British Museum, Herakles sacrifices to Chryse in her precinct, in which are to be seen a statue of the goddess on a column, an altar, and a leafy tree from which three votive tablets or *pinakes* dangle (Figure 1.1).[54] The sacrificial animal's tail, the god's portion of the sacrifice, burns on the fire, while the skewered meat to be consumed by the worshippers is roasted over it.

The single tree standing *pars pro toto* as an artistic convention for an entire grove is a feature also of marble votive reliefs that were thank offerings to the gods in fulfilment of a vow. On a marble relief panel in the Glyptothek in Munich, dating to about 200–150 BC, a group of eight men, women, and children approaches a bearded god and his female associate, bringing objects of dedication and sacrifice (Figure 1.2).[55] They all are gathered in a

Figure 1.1 Athenian red-figure bell *krater* with worshippers sacrificing in a sacred grove, c. 450–425 BC (I. Deluis).

Figure 1.2 Greek votive relief depicting worshippers in a sacred grove approaching the gods, c. 200–150 BC (Hartwig Coppermann, courtesy of the Glyptothek, Munich).

sanctuary planted with trees, although the grove here is represented as a single, old plane tree with leafy, gnarled branches. The sanctity of the tree and the place in which it stands is obvious because of the sacrificial altar below it, as well as the dedicated cloth fillets wrapped around the tree trunk, and the tall pedestal on which small statues of two gods stand. A curtain tied to a branch of the tree on the left and another tree or structure out of the picture separates the gods and their worshippers from the rest of the open-air sanctuary. The antiquity of the sacred grove is particularly highlighted by the apparent age and size of the plane tree.

A popular genre of Roman wall painting in the late first century BC and the first third of the first century AD was that of the so-called sacro-idyllic landscape. These landscapes, executed in an exquisite and impressionistic, even sketchy fashion, were painted on the walls of private houses in Roman Italy, with the best and most complete examples surviving in Rome and in and around Pompeii (Figure 1.3). Some of the most beautiful landscapes of this type decorated the imperial villa of Augustus's friend and son-in-law Agrippa at Boscotrecase near Pompeii.[56] In these paintings, rural shrines consisting of a few remains of ruined buildings, a tower or a votive column

Figure 1.3 Roman wall painting from the Villa Farnesina, Rome, depicting a rural
shrine with trees and a cult statue on a column, late first century BC
(M. Carroll).

are located on rocky outcrops, and a few gnarled trees casting a bit of shade
on the tiny figures of humans and animals moving around in and around
the sanctuary. They are very much of their age when the bucolic landscape
of streams, meadows, groves, and hills peopled by goatherds, shepherds,
fauns, satyrs, and nymphs became a major literary genre in the poetry of the
Augustan age, primarily due to Virgil's *Eclogues*.[57] For city dwellers, these
idealised and pleasant, if artificial, landscapes offered a vision of a benign
pastoral world in which the natural environment was revered but nonetheless tamed and to which they symbolically had access. Urban and rural ways
of life blended together in these paintings, as they did in reality, for example, at the sacred grove (*nemus*) of Anna Perenna just outside Rome. Ovid
described the use of the site by the common people of Rome, who came
on her annual festival on 15 March to have picnics on the meadows, some
pitching tents, some making leafy huts of branches or awnings of togas
draped over reed stalks, and the description sounds a lot like the painted
bucolic landscapes of age-old sacred groves.[58] This sanctuary, however, was
neither ancient nor natural in character. Archaeological exploration of the
deposits in a sacred well at the grove shed light on the votive offerings left

behind by the faithful.[59] They indicate that the main periods of activity here cannot be ascribed to the distant past, but to the Augustan period and the fourth century AD. This sanctuary and its grove were in a suburban area on the border of the city of Rome, not in a rural location, and its visitors were city dwellers, not countryside rustics, who could enjoy a religious experience in accessible and managed nature.[60]

Temples and sacred groves occasionally are depicted also on Roman coins, particularly those from various places in the Greek-speaking Eastern Empire. Bronze coins of Caracalla (AD 198–217) and Elagabalus (AD 218–222) from Thrace, for example, show the emperor's portrait on one side, and on the reverse is a portrayal of a temple at Augusta Traiana (Stara Zagora) in modern Bulgaria.[61] On either side of the temple is a tree, suggesting that there was a grove surrounding it. A temple is also on coins from Zeugma (Belkis) on the Euphrates, issued in the second and third centuries AD. In front of the temple is a colonnaded square or *porticus* with a large grove of trees within it.[62]

Excavated remains of sacred groves

There is a considerable body of archaeological evidence for sacred groves in sanctuaries. Plantings of trees in the pilgrimage sanctuary of Apollo Hylates, or 'Apollo of the woods', at Kourion on the south coast of Cyprus, for example, are known both from textual and archaeological evidence.[63] The excavations demonstrated that trees or bushes were planted in the seventh or sixth century BC in pits and trenches cut into the bedrock in the sanctuary *temenos*, although the grove went out of use and the planting pits were paved over by the first century BC. But it is possible that the cult also owned woodlands separate from the enclosed temple site itself, especially since Apollo was revered at Kourion as the god of woodlands. It may be these woodlands that are described much later in the second century AD as an extensive grove filled with wild animals, primarily deer.[64]

There is no literary or epigraphic mention of a sacred grove associated with the temple of Hephaistos in the heart of Athens, but archaeological excavations in 1937 uncovered rows of planting pits arranged around three sides of the temple.[65] The pits were cut out of the living rock of the slope on which the temple stood, some of the pits in 1937 still containing ceramic planters for trees or bushes (Figure 1.4). The diagnostic finds and stratigraphy of the deposits indicate that although the temple was built in the mid-fifth century BC, the trees around it were not planted until about two centuries later. This grove may have been damaged in 86 BC in the Roman siege of Athens under Sulla, as the trees were replanted thereafter in the first century BC.

In Italy, one of the earliest examples of a Roman temple grove is that of Juno at Gabii near Rome, dating to the second half of the second century BC.[66] Excavations at the site revealed rows of thirty-four holes (1.50 x 1.60 m) cut into the bedrock for trees; these were arranged on three sides of the temple, much like the plantings around the temple of Hephaistos in Athens.

Figure 1.4 Excavated planting pits for trees cut into the living rock next to the temple of Hephaistos, Athens (American School of Classical Studies).

In the first decades of the first century BC, the grove was restructured and seventy new and smaller holes (1.20 x 1.30 m) were excavated for plantings. These may have been for less substantial trees or bushes than those in the first phase, possibly myrtles, although no botanical remains have survived to confirm this.

Evidence for the planting of trees and shrubs in temple precincts in the city of Rome itself is varied. Some of it is restricted solely to written sources.

A shrine, known only from inscriptions, is that of Bellona Pulvinensis in the north-east part of the city. A funerary inscription of a priest of Bellona from Rome refers to the shrine as having been within a grove (*in luco*).[67] There was another grove in the sanctuary of Libitina. Dionysius of Halicarnassus referred to a 'treasury of Venus in the Grove, whom they call Libitina'.[68] The *lucus Libitinae* may have been located just outside the Esquiline gate in the eastern part of the city where there are attested burial grounds, although funerary inscriptions naming individuals who lived 'near the grove of Libitinae' do not reveal so precise a location.[69] John Bodel, who studied a range of texts in conjunction with the grove of Libitina in Rome, concludes that in this location the business of professional undertaking was conducted, a death register kept, and a possible death tax paid.[70] This is, therefore, not a typical sacred grove.

Archaeological exploration in Rome also has revealed physical remains of sacred groves. The remains of soil and roots embedded in rows of travertine marble containers on the west and north sides of the temple of the deified Julius Caesar, built in the Roman forum in 29 BC, for example, indicate that these containers held plantings, and the organic remains point to laurel trees.[71] In the early 1990s, excavations by the *École française* at a site on the Palatine hill uncovered the remains of a system of plantings within a complex that may have belonged to the eastern god Elagabal erected by the emperor Elagabalus in the early third century AD.[72] Surrounded by porticoes on all four sides, the temple stood in the centre of a courtyard that was paved, but the paving was interrupted by four rectangular planting beds irrigated by underground channels. Although it is not certain what type of plants grew in the courtyard, low plantings of bushes and small trees have been suggested. The pots used for plantings were halved amphorae, the majority of which are wine and oil amphorae of the late second and early third century AD (Figure 1.5). The discovery of broken amphorae re-used as planting pots at this and other sites makes it apparent that such containers often had a 'second life' in the context of both secular and sacred gardens.[73]

A marble plan of the city of Rome carved in the reign of Septimius Severus at the beginning of the third century AD, known as the *Forma Urbis Romae*, is another valuable source of information not only on the buildings in Rome, but also on groves and gardens associated with several structures.[74] The excavated and surviving fragments of the *Forma Urbis*, together with archaeological exploration of the sites depicted on the plan, clearly show that formal plantings were integral design elements in the temples of Rome. These plantings are represented in a variety of ways on the marble plan. The building complex erected in 55 BC in the western suburb of Rome by Pompey the Great to celebrate his triple triumph over Mediterranean pirates and kingdoms in Asia Minor and the Black Sea in 61 BC included a stone-built theatre and a temple of Venus adjacent to a large colonnaded courtyard (the *Porticus Pompeiana*). On the marble plan, two elongated rectangles parallel to the two long sides of the courtyard can be

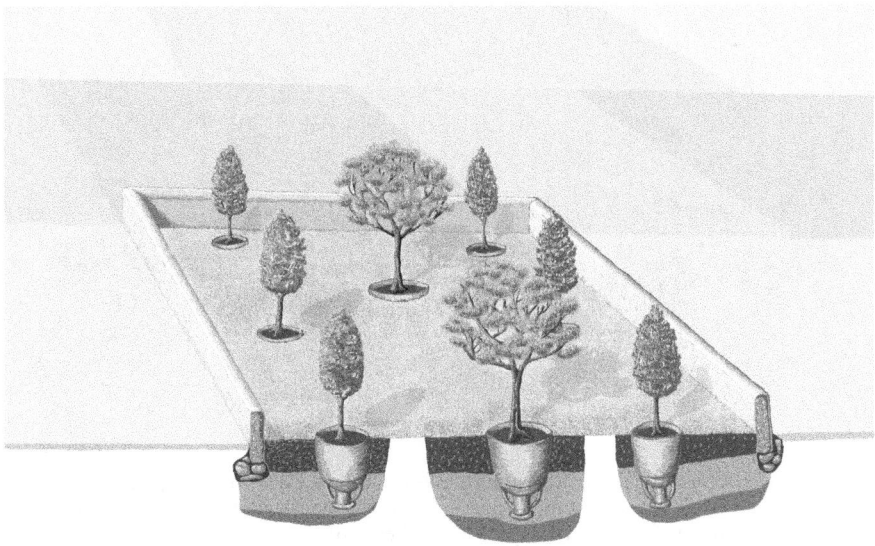

Figure 1.5 Reconstruction of an area of the sacred grove at the Temple of Elagabal in Rome, where the amphorae (upside down) were used as planters in the third century AD (I. Deluis).

seen. Gleason's definitive landscape architectural study of this complex has shown that these rectangles represent the double lanes of plane trees praised in ancient descriptions of the building, perhaps the Asiatic plane trees that Pompey brought back from his campaigns and displayed in his triumph as fruits of victory (Figure 1.6).[75] The poet Martial, writing in the first century AD, referred to the plantings here as a *nemus duplex*, a double grove, the term *nemus* indicating the sacred nature of this carefully contained designed landscape.[76] Between the columns of the porticoes hung gold embroidered tapestries from Pergamon, and in the galleries and the grove were famous masterpieces of painting, statues of gods, and mythical figures (some with fountains), as well as images of conquered nations, all impressing members of the Roman public who strolled here as testament to the wealth of Roman conquest and the divine sanction of Pompey and Rome by the goddess Venus.[77]

The marble plan also shows rows of joined rectangles in the courtyard of the *Templum Pacis* (Temple of Peace) in the heart of Rome, dedicated by the emperor Vespasian in AD 75 in celebration of his victory in Judea. These once were thought to represent rows of trees, hedges, or flowerbeds, but recent excavation at the site indicates that these were, in fact, masonry water basins and bases for sculpture.[78] Between the water basins were plantings,

Figure 1.6 Reconstruction of the *nemus duplex* in the Porticus Pompeiana in Rome with the theatre/temple of Venus in the background, 55 BC (L. Cocker-ham Catalano, courtesy of K.L. Gleason).

as indicated by the double rows of amphorae found in the soil that had been re-used as planting pots. It is unknown what kinds of trees or bushes once grew in these pots, but at least some of the vegetation consisted of rose bushes, to which the carbonised remains of this plant in the soil attest. Like the *Porticus Pompeiana*, the galleries and courtyard here were adorned with famous Greek masterpieces of painting and sculpture, and in this case also trophies from Jerusalem.[79] Although the building was also called a forum, which seems to have little to do with a sanctuary, other Latin and Greek authors call the overall site *templum* and *aedes* (temple and shrine) as well as *temenos* and *hieron* (sacred precinct and sanctuary).[80] We do not know whether the plantings in the courtyard were called a *nemus*, but the presence of a temple to *Pax* (peace) suggests that it had a sacred character, in addition to its role as a display of Roman conquest and dominion over others.[81]

A temple grove of the first century BC is attested also at Pompeii. Excavations by the author at the temple of Venus, the patron goddess of the city, produced clear evidence that this sanctuary was built in the mid-first century BC after Pompeii had become a Roman colony under Sulla (Figure 1.7).[82] At the same time, the courtyard surrounding the temple was planted to create a sacred grove. The goddess of fertility, Venus, was connected more than any other with vegetation and growing plants, so it is entirely appropriate that her precinct should be planted. On three sides of the courtyard in front of the porticoes, planting pits, terracotta plant pots, and root cavities were found. The pits contained ceramic pots in a complete or fragmentary state; these vessels alternated between those either with one hole in the base or with one hole in the base and three in the lower body, and in between

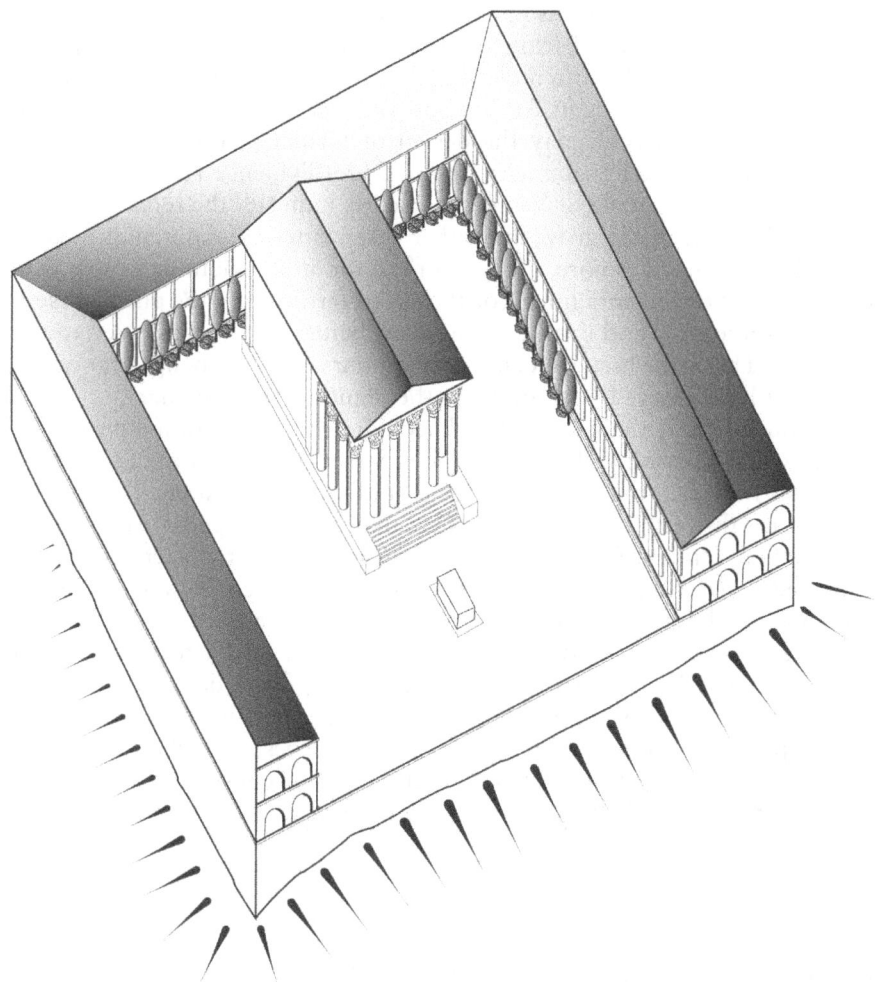

Figure 1.7 Plan of the sanctuary of Venus at Pompeii with tree plantings in the grove
on three sides of the temple courtyard, ca. 50 BC (O. Jessop).

these were pits with no pot in them at all. This regular pattern may reflect
the planting of types of vegetation of alternating form and shape, or it may
simply imply a sequence of trees and shrubs that required propagation in
ceramic pots and those that did not need containers to develop. Possibly
laurels, roses, and myrtles grew in these containers, the latter two being
particularly sacred to Venus.

The sacred grove of Venus, almost certainly a *nemus*, was very much an
architectural garden in which alternating types of trees and bushes were

planted parallel to the columns of the porticoes surrounding the courtyard, the trees echoing the rhythms of the columns and visually highlighting the temple as one approached from the south. In its layout of regular plantings on three sides of the temple the grove of Venus resembled those at the temple of Hephaistos in Athens and the temple of Juno at Gabii (see above). All three sites display the design of formal plantings surrounded by a *porticus triplex* in sanctuary architecture. Evidence recovered in the excavations at the temple of Venus also clearly indicates the replacement of some of the trees in the grove and other rarely attested maintenance work by gardeners.[83] Furthermore, analysis of the soil in the planting pots points to the practice of soil enrichment or the use of fertilisers in the plant nursery in which these trees and bushes were raised before being transferred to the sanctuary. This is the first evidence for plant nurseries supplying cult organisations with the desired vegetation for the planting of a sanctuary.

Also outside Italy there is good archaeological evidence for temple groves of the Roman period.[84] The geographer Strabo referred to the grove of laurels surrounding the gymnasium and stadium at Nikopolis, the Roman town in north-west Greece that was created by the emperor Augustus in the late first century BC.[85] But this emperor also built a sacred precinct dedicated to Mars, Neptune, and Apollo here between 29 and 27 BC to celebrate his defeat of Antony and Cleopatra at the battle of Actium in 31. Recent excavations show that the courtyard in this victory monument was a formally planted area within a *porticus triplex*, a design that clearly had become popular beyond Italy.[86] Ceramic plant pots were inserted in the ground parallel to and in front of all three porticoes, with the altar in the centre of the courtyard. Augustus himself must have chosen the trees for his monument, and it is likely that they were laurels, as these were particularly sacred to Apollo, his patron god.

Conclusion

Throughout the Greek and Roman worlds, 'the sacred groves of the immortal ones' were enclosed or assigned spaces, separate from secular ones, where mortals worshipped and communicated with the divine. Particularly revered were those sacred groves that were perceived to be of great antiquity and unmanaged, although, in reality, the vast majority of open-air cult sites were generally shaped by man, either by enclosing them, setting up altars and votive monuments and building temples in them, or even planting a regular and regimented grove of trees from scratch. From the first century BC, and especially in Roman Italy, many completely new temples and sanctuaries were established that included the recurring design of rows of trees on two or three sides of a sacred precinct, ranging from the *nemus duplex* of Pompey the Great in Rome to the *porticux triplex* at the temple of Venus in Pompeii. These have the appearance of being manicured and managed in the extreme. At the same time, and particularly in the Augustan age, wall paint-ing and poetry revelled in the bucolic portrayal of apparently ancient, rural landscapes in which little shrines and shady groves are visited by shepherds and other rustics, but the natural environment in such portrayals is artificial and, for increasing numbers of city dwellers, primarily symbolic of a benign pastoral world of old in which gods and men interacted in tune with nature.

Notes

1 For a 'garden for the god Addu' that was 'full of juniper trees' in the late nine-teenth and early eighteenth centuries BC, see Jean-Claude Margueron, 'Die Gärten im Vorderen Orient', in *Der Garten von der Antike bis zum Mittelalter*, ed. by Maureen Carroll-Spillecke (Mainz: Verlag von Zabern, 1992), p. 61. On Egyptian gardens and sacred gardens, see Jean-Claude Hugonot, *Le jardin dans l'Egypte ancienne* (Frankfurt: Peter Lang, 1989); Alix Wilkinson, *The Garden in Ancient Egypt* (London: Rubicon Press, 1998). *Isaiah* 60.13 records trees planted to beautify the sanctuary of the Hebrew god.

2 Lines 256–72. The poem probably dates to the period between 700–500 BC. It is translated and commented on by Diane Raynor, *The Homeric Hymns: A Trans-lation With Introduction and Notes* (Berkeley: University of California Press, 2004), pp. 75–85. See also Borimir Jordan and John Perlin, 'On the protection of Sacred Groves', in *Studies Presented to Sterling Dow on His Eightieth Birthday*, ed. by Alan L. Boegehold, et al (Durham, NC: Duke University Press, 1984), p. 154.

3 For the garden of the Hesperides in the Atlas Mountains, in which the goddess Hera had planted the tree of immortality laden with golden apples, as well as other trees such as pomegranate, pear, mulberry, myrtle, laurel, and almond, see Hesiod, *Theogony* 215–16 and Skylax, *Periplus* 108.

4 Dunstan Lowe, 'Tree-worship, Sacred Groves and Roman antiquities in the Aeneid', *Proceedings of the Virgil Society* 27:1 (2011), pp. 91–128.

5 Pausanias, *Description of Greece* 8.54.7. See also Christian Jacob, 'Paysage et bois sacré: ἄλσος dans la *Périégèse de la Grèce* de Pausanias', in *Les Bois Sacrés*, ed. by Olivier de Cazanove and John Scheid (Naples: Centre Jean Bérard, 1993), pp. 31–44. On the portrayal of trees and ruins by Pausanias, see James I. Porter, 'Ideals and ruins: Pausanias, Longinus and the second sophistic', in *Pausanias: Travel and memory in Roman Greece*, ed. by Susan E. Alcock, John F. Cherry and Jas Elsner (Oxford: Oxford University Press, 2001), pp. 63–92.

6 *IG* 12.1.781–3; Georgios Deligiannakis, 'Late paganism on the Aegean Islands and processes of Christianisation', in *The Archaeology of Late Antique Pagan-ism*, ed. by Luke Lavan and Michael Mulryan (Leiden: Brill, 2011), pp. 314–15.

7 Janet Grossman, *Greek Funerary Sculpture. Catalogue of the Collections at the Getty Villa* (Los Angeles: Getty Publications, 2001), cat. No. 47, pp. 130–31.

8 Stephen G. Miller, 'Excavations at Nemea 1976,' *Hesperia* 46 (1977), pp. 1–26 and 'Excavations at Nemea 1977', *Hesperia* 47 (1978), pp. 58–88; Darice E. Birge, Lynn H. Kraynak and Stephen G. Miller, *Excavations at Nemea: Top-ographical and Architectural Studies. The Sacred Square, the Xenon and the Bath* (Berkeley: University of California Press, 1992), pp. 85–96 and figs. 1, pp. 98–103.

9 Pausanias, *Description of Greece* 2.15.2

10 See Philippe Bonnachere, 'The place of the Sacred Grove (alsos) in the Mantic rituals of Greece: The example of the alsos of Trophonios at Lebadeia (Boiotia)',

in *Sacred Gardens and Landscapes: Ritual and Agency*, ed. by Michel Conan (Washington, DC: Dumbarton Oaks, 2007), pp. 17–18.

11 On ancient Greek gardens of all kinds, see Maureen Carroll-Spillecke, *ΚΗΠΟΣ. Der antike griechische Garten* (Munich: Deutscher Kunstverlag, 1989) and *Der Garten von der Antike bis zum Mittelalter* (Mainz: Verlag von Zabern, 1992), pp. 153–76.

12 John Scheid, 'Lucus, nemus. Qu'est-ce qu'un bois sacré?', in *Les Bois Sacrés*, ed. by Olivier de Cazanove and John Scheid (Naples: Centre Jean Bérard, 1993) p. 19; Filippo Coarelli, 'I luci del Lazio: la documentazione archeologica', in *Les Bois Sacrés*, ed. by Olivier de Cazanove and John Scheid (Naples: Centre Jean Bérard, 1993), p. 52.

13 Ingrid E. M. Edlund, *The Gods and the place: Location and function of sanctuaries in the countryside of Etruria and Magna Graecia (700–400 B.C.)* (Stockholm: Almqvist and Wiksell, 1987), pp. 30–43, pp. 126–46.

14 *CIL* 6.32455/*ILS* 5428.

15 Pausanias, *Description of Greece* 8.37.10, 2.27.1.

16 Ulrike Egelhauf-Gaiser, 'Roman cult sites: A pragmatic approach', in *A Companion to Roman Religion*, ed. by Jörg Rüpke (Oxford: Blackwell, 2007), p. 206.

17 Livy, *From the Founding of the City* 24.3.3–7. The cattle were not merely for decoration, however, as Livy tells us that 'great profits were made from the cattle', and out of these profits a massive golden column was made and set up to the goddess.

18 Jordan and Perlin, pp. 155–7.

19 Jordan and Perlin, p. 156.

20 John Bodel, *Graveyards and Groves: A Study of the Lex Lucerina* (Cambridge, MA: American Journal of Ancient History, 1994), p. 26.

21 *CIL* 6.4766–4767/*ILS* 4911. Bodel, pp. 24–9.

22 Valerius Maximus, *Memorable Doings and Sayings* 1.19; Cassius Dio, *Roman History* 51.8.2.

23 Polybios, *Histories* 16.1.6.

24 Jean Delorme, *Gymnasion: Étude sur les monuments consacrés à l'éducation en Grèce* (Paris: Boccard, 1960), pp. 51–61; John Travlos, *Pictorial Dictionary of Ancient Athens* (London: Thames and Hudson, 1971), pp. 42–51, 340–41, 345–47; Carroll-Spillecke, *ΚΗΠΟΣ*, pp. 28–31; Wolfram Hoepfner, *Antike Bibliotheken* (Mainz: Verlag von Zabern, 2002), pp. 56–62; Maureen Carroll, *Earthly Paradises: Ancient Gardens in History and Archaeology* (London: British Museum Press, 1993), pp. 50–2. For the intentional planting of trees in these areas, see Plutarch, *Cimon* 13; Pseudo-Plutarch, *Decem Oratorum Vitae* 841C-D.

25 Aristophanes, *Clouds* 1002–1019; Plato, *Laws* 6.761B-C; Diogenes Laertius, *Lives of Eminent Philosophers* 3.7.

26 Plutarch, *Sulla* 12.3; Appian, *Mithridatic Wars* 30.

27 *Civil War* 3.399–425.

28 Bonnachere, pp. 1–7.

29 Vitruvius, *On Architecture* 4.8.4. For an overview of the archaeology of the sanctuary, see Pia Guldager Bilde and Mette Moltesen, *In the Sacred Grove of Diana: Finds From the Sanctuary at Nemi* (Copenhagen: Ny Carlsberg Glyptothek, 1997).

30 Strabo, *Geography* 5.3.12; Virgil, *Aeneid*, 6.13; Servius, *On Aeneid* 6.136; Suetonius, *Caligula* 35; Pausanias, *Description of Greece* 2.27.4.

31 Michael H. Crawford, *Roman Republican Coinage* (Cambridge: Cambridge University Press, 2001), no. 486/1.

32 Ovid, *Fasti* 3.266.

33 *CIL* 6.2065; *CIL* 6.2075; Henri Broise and John Scheid, 'Étude d'un cas: Le lucus Deae Diaea à Rome', in *Les Bois Sacrés*, ed. by Olivier de Cazanove and John Scheid (Naples: Centre Jean Bérard, 1993), pp. 145–57.
34 Pliny, *Letters* 8.8.
35 Suetonius, *Caligula* 43.
36 *Natural History* 12.2.3.
37 Livy, *From the Founding of the City* 10.23.12; Pliny, *Natural History* 15.20.77.
38 Pliny, *Natural History* 15.36.120–1.
39 *Metamorphosis* 14.836–7.
40 Pliny, *Natural History* 16.57.132.On the wars against the Germanic Cimbri and Teutones, see Maureen Carroll, 'Measuring time and inventing histories in the early empire: Roman and Germanic perspectives', in *TRAC 2001: Proceedings of the Eleventh Annual Theoretical Roman Archaeology Conference, Glasgow 2001*, ed. by Martin Carruthers (Oxford: Oxbow, 2002), pp. 104–12.
41 Herodotos, *History* 6.108.4.
42 Diodorus Siculus, *Historical Library* 12.39.1; Plutarch, *Pericles* 3.2; Lycurgus, *Against Leocrates* 93.
43 Statius, *Thebaid* 481–296; Homer A. Thompson and R. E. Wycherley, *The Agora of Athens. The History, Shape and Uses of an Ancient City Center* (Princeton: Princeton University Press, 1972), p. 135; L. M. Gadberry, 'The Sanctuary of the Twelve Gods in the Athenian Agora: A Revised View', *Hesperia* 61 (1992), pp.447–89.
44 *Historical Library* 5.42.6–43.3, 5.44.5
45 The map is known today in a medieval copy, the so-called *Tabula Peutingeriana*. Annalina and Mauro Levi, *Itineraria picta: Contributo allo studio della Tabula Peutingeriana* (Rome: 'L'Erma' di Bretschneider, 1967), pp. 154–56, pl. 9; Carroll 1993, 71, fig. 55. On primary references to the grove, see Strabo, *Geography*, 16.2.6; Philostratos, *Life of Apollonios* 1.16.
46 *IG* I³ 84; Travlos, pp. 332–33, fig.435.
47 Susan G. Cole, *Landscapes, Gender, and Ritual Space: The Ancient Greek Experience* (Berkeley: University of California Press, 2004), pp. 59–60.
48 John Harvey Kent, 'The Temple Estates of Delos, Rheneia and Mykonos', *Hesperia* 17 (1948), pp. 243–338.
49 *IG* XII.7.62.
50 *IG* 14.645
51 Folkert T. van Straten, *Hiera Kala: Images of Animal Sacrifice in Archaic and Classical Greece* (Leiden: E.J. Brill, 1995), p. 215, cat. No. V127, fig.30.
52 van Straten, p. 231, cat. no. V200, fig.152.
53 On animal sacrifice and the apportioning of sacrificial meat to man and gods, see van Straten, pp. 128–41.
54 van Straten, p. 261, cat. no. V367, fig.124.
55 Jerome J. Pollitt, *Art in the Hellenistic Age* (Cambridge: Cambridge University Press, 1986), p. 197, fig.210.
56 Peter von Blanckenhagen and Christine Alexander, *The Augustan Villa at Boscotrecase* (Mainz: Philipp von Zabern, 1990). For bucolic rural shrines, see also the paintings in the Villa Farnesina in Rome: Diana Spencer, *Roman Landscape: Cultural and Identity* (Cambridge: Cambridge University Press, 2010), pp. 147–53, figs.13–14.
57 Frederick Jones, *Virgil's Garden: The Nature of Bucolic Space* (London: Bloomsbury, 2013), pp. 18–21.
58 Ovid, *Fasti* 3.525–30. See also Martial, *Epigrams* 4.64.17.
59 Marina Piranomonte, *Il santuario della Musica e il bosco sacre di Anna Perenna* (Milan: Electa, 2002).
60 Egelhauf-Gaiser, pp. 213–14.

61 Nikola Muschmov, *Antichnitje moneti na Balkanskiia poluostrov i moneti-tie tsare (Ancient Coins of the Balkan Peninsula and the Coins of the Bulgarian Monarchs)* (Sofia: National Museum, 1912), cat. no. 3076; Richard Stoll, *Architektur auf römischen Münzen.* (Trier: Selbstverlag, 2000), p. 80, nos. 115–16.

62 Warwick Wroth, *A Catalog of the Greek Coins in the British Museum, Galatia, Cappadocia, and Syria* (London: British Museum, 1899), p. 127, no. 29, pl.16.13, no.46, pl.16.14.

63 Darice E. Birge, 'Ἄλσος and the Sanctuary of Apollo Hylates', in *Studies in Cypriote Archaeology*, ed. by Jane C. Biers and David Soren (Los Angeles: Institute of Archaeology, 1981), pp. 153–58; Vassos Karageorghis and Maureen Carroll-Spillecke, 'Die heiligen Haine und Gärten Zyperns', in *Der Garten von der Antike bis zum Mittelalter*, ed. by Maureen Carroll-Spillecke (Mainz: Verlag von Zabern, 1992), pp. 141–52.

64 Aelian, *Nature of Animals* 11.7.

65 Dorothy B. Thompson, 'The Garden of Hephaistos', *Hesperia* 6 (1937), pp. 396–425; John M. Camp, *The Athenian Agora: Excavations in the Heart of Classical Athens* (London: Thames and Hudson, 1986), p. 87; Carroll-Spillecke, *ΚΗΠΟΣ*, pp. 31, 58, 72.

66 Hans Lauter, 'Ein Tempelgarten?', *Archäologischer Anzeiger* (1968–69), pp. 626–31; Coarelli, pp. 48–51.

67 *CIL* 6.2232/*ILS* 4181.

68 *Roman Antiquities* 4.15.5.

69 *CIL* 6.9974/*ILS* 7574; *CIL* 6.33870/*ILS* 7471.

70 Bodel, pp. 17–18.

71 Bernhard Andreae, 'Archäologische Funde im Bereich von Rom 1949–1956–57', *Archäologischer Anzeiger* (1957), pp. 110–358; Lawrence Richardson, *A New Topographical Dictionary of Ancient Rome* (Baltimore: Johns Hopkins University Press, 1992), pp. 213–14.

72 Françoise Villedieu, 'I giardini del tempio', in *Il giardino dei Cesari: Dai palazzi antichi alla Vigna Barberini sul monte Palatino. Scavi dell'École Française de Rome, 1985–1999*, ed. by Françoise Villedieu (Rome: Edizioni Quasar, 2001), pp. 94–7; Giorgio Rizzo, 'Le anfore del giardino del tempio', in *Il giardino dei Cesari: Dai palazzi antichi alla Vigna Barberini sul monte Palatino. Scavi dell'École Française de Rome, 1985–1999*, ed. by Françoise Villedieu (Rome: Edizioni Quasar, 2001), p. 98.

73 For example, at Hadrian's villa at Tivoli: Wilhelmina F. Jashemski and Eugenia S. P. Ricotti, 'Preliminary Excavations in the Gardens of Hadrian's Villa: The Canopus Area and the Piazza d'Oro', *American Journal of Archaeology* 96 (1992), pp. 580–85.

74 Gianfilippo Carettoni, Antonio M. Colini, Lucos Cozza, and Guglielmo Gatti, *La pianta marmorea di Roma antica* (Rome: Comune di Roma, 1960); Robert B. Lloyd, 'Three Monumental Gardens on the Marble Plan', *American Journal of Archaeology* 86 (1982), pp. 91–100.

75 Kathryn Gleason, 'Porticus Pompeiana: A new perspective on the first public park of ancient Rome', *Journal of Garden History*, 14.1 (1994), pp. 13–27. For a description of the living trees from Asia and Africa that were brought in Pompey's triumphal procession, see Pliny the Elder, *Natural History* 12.20 and 12.111.

76 Martial, *Epigrams* 2.14.10. The trees (and their shade) are mentioned again by Martial, *Epigrams* 11.47.3 and Propertius 2.32.11–12.

77 Ann L. Kuttner, 'Culture and history at Pompey's museum', *Transactions of the American Philological Association* 129 (1999), pp. 343–73.

78 Carettoni *et al.*, pp. 73–4, pl. 20.15–16; Lloyd, pp. 91–2, fig. 1; Richardson, pp. 286–87; S. Rizzo, 'Indagini nei fori Imperiali. Oroidrografia, foro di Cesare, foro di Augusto, templum Pacis', *Römische Mitteilungen* 108 (2001), pp. 238–39; Roberto Meneghini, 'I Fori Imperiali: ipotesi ricostruttive ed evidenze archeologica', in *Imaging Ancient Rome. Documentation – Visualisation – Imagination*, ed. by Lothar Haselberger and John Humphrey (Portsmouth, RI: Journal of Roman Archaeology, 2006), pp. 158–59; Stefania Fogagnolo, 'Lo Scavo del Templum Pacis: Concordanze e novità rispetto alla *Forma Urbis*', in *Formae Urbis Romae: Nuovi Frammenti di Piante Marmoree dallo Scavo dei Fori Imperiali*, ed. by Roberto Meneghini and Riccardo Santangeli Valenzani (Rome: 'L'Erma' di Bretschneider, 2006), pp. 61–74; Roberto Meneghini and Riccardo Santangeli Valenzani, *I Fori Imperiali. Gli Scavi del Comune di Roma (1991–2007)* (Rome: Viviani Editore, 2007), pp. 61–70, figs.54–5. On the use of exotic trees and plants brought back to Rome in triumph over defeated lands, including the parading of balsam trees from Judaea by Vespasian in AD 71, see Elizabeth Macaulay-Lewis, 'The Fruits of Victory: Generals, Plants and Power in the Roman World', in *Beyond the Battlefields. New Perspectives on Warfare and Society in the Graeco-Roman World*, ed. by Edward Bragg, Lisa I. Hau and Elizabeth Macaulay-Lewis (Cambridge: Cambridge Scholars Publishing, 2008), pp. 206–24.

79 Josephus, *The Jewish War* 7.5.7; Pliny the Elder, *Natural History* 12.94; 34.84; 35.102, 109; 36.27, 58; Suetonius, *Vespasian* 18. Some of the sculpture had been taken from the demolished Golden House of Nero.

80 Carlos F. Noreña, "Medium and message in Vespasian's Templum Pacis", *Memoirs of the American Academy in Rome* 48 (2003), pp. 25–6, 40.

81 Noreña, pp. 38–41.

82 Maureen Carroll, 'Nemus et Templum. Exploring the sacred grove at the Temple of Venus in Pompeii', in *Nuove ricerche archeologiche nell'area vesuviana (scavi 2003–2006), Atti del convegno internazionale, Roma 1–3 Febbraio 2007*, ed. by Pietro Giovanni Guzzo and Maria Paola Guidobaldi (Rome: 'L'Erma' di Bretschneider, 2008), pp. 37–45, and 'Exploring the sanctuary of venus and its Sacred Grove: Politics, cult and identity in Roman Pompeii', *Papers of the British School at Rome* 78 (2010), pp. 63–106.

83 Carroll 2010, pp. 79–80.

84 These include tree pits in a temple precinct in Thuburbo Maius in Tunisia and planting pits in the paved terrace of a temple complex at Munigua in Spain. Wilhelmina F. Jashemski, 'Roman gardens in Tunisia: Preliminary excavations in the House of Bacchus and Ariadne and in the East Temple at Thuburbo Maius', *American Journal of Archaeology*, 99 (1995), pp. 559–76; Theodor Hauschild, 'Los templos romanos de Munigua', in *Templos romanos de Hispania* (Murcia: Cuadernos de arquitectura romana, 1992), pp. 133–43; Coarelli, p. 52.

85 *Geography* 7.7.6.

86 Konstantin Zachos, 'The tropaeum of the sea-battle of Actium at Nikopolis: Interim report', *Journal of Roman Archaeology* 16 (2003), pp. 65–92.

2 Seeing the wood for the trees
The long-term aesthetics of woodland in England

Tom Williamson

Introduction: groves and wildernesses

Groves, wildernesses and other forms of ornamental woodland, dissected by paths and clearings and containing a range of buildings, sculptures and other features, were popular elements in English gardens in the later seventeenth and early eighteenth centuries.[1] They came in a bewildering variety of forms, which are discussed in more detail by other contributors to this volume; and they changed significantly over time, both in form and location. Put perhaps over-simply, they gradually became a more prominent element in large gardens in the decades following the Restoration so that by the 1720s they often took up most of their area; and they came to include not only straight but also serpentine walks or allées. Most originally appear to have had hedged paths and to have contained coppiced underwood, but by the late 1720s 'open' groves, in which the paths lacked flanking hedges and in which there was often no coppiced understorey, were increasing in importance.[2] There was also an apparent tendency for areas of detached woodland garden, 'forest gardens' to some writers, to become more popular, lying at a distance from a mansion, often on the far side of a park (Figures 2.1 and 2.2).

Although garden historians quite rightly emphasise the importance of these varied forms of ornamental wood or woodland garden in the period after the Restoration, they were present in some form in many gardens at a rather earlier date, and we should always be careful not to fall into that old trap, of dating the origins of particular garden features to the time that they first appear clearly in our sources. Trees take time to grow. Mature woods shown on illustrations made by Johannes Kip and Leonard Knyff around 1700 were often significantly more than four decades old. There are, we should note, far fewer maps and illustrations of country houses and their grounds dating to the period before the Restoration, but where these exist they often show very similar features: Figure 2.3, for example, shows a detail from a map of 1652 depicting the wilderness in the gardens at Somerleyton in Suffolk, which other sources suggest was planted in the 1620s.

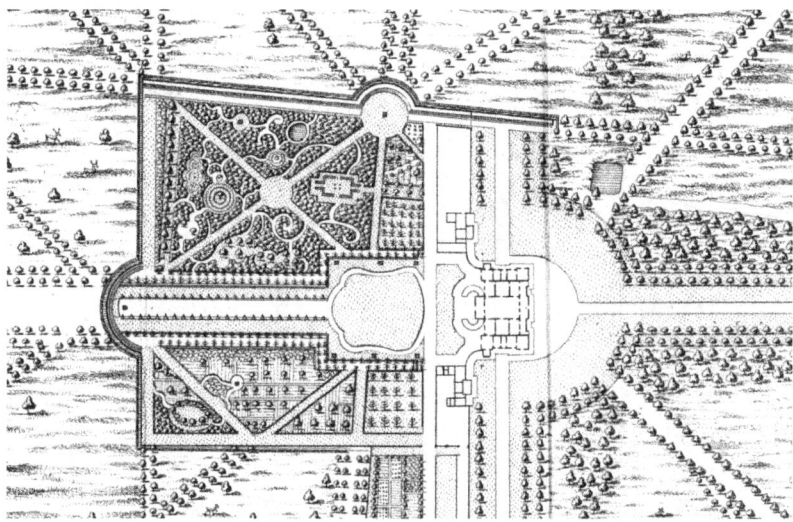

Figure 2.1 Houghton in Norfolk, as depicted in Colen Campbell's *Vitruvius Britannicus* of 1725. The 'wilderness' to the north of the main east-west vista typically takes up a substantial proportion of the garden area (Tom Williamson).

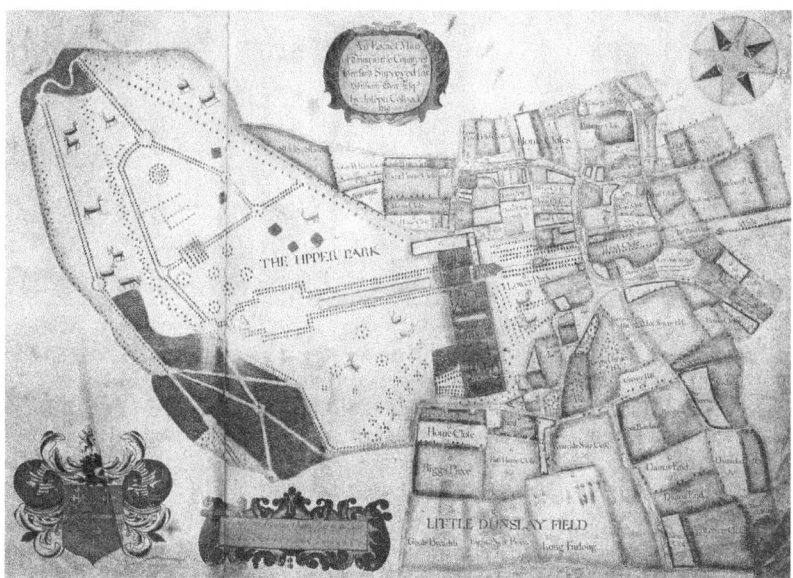

Figure 2.2 Tring Park, Hertfordshire, as shown on a map dated 1719, but added to and revised on subsequent occasions. The woodland garden or detached wilderness created by Charles Bridgeman can be seen bottom left, with obelisk and temple designed by James Gibbs. It occupied steeply rising ground on the Chiltern escarpment, on the other side of the park from the mansion itself (private collection).

Figure 2.3 Somerleyton Hall, Suffolk, as depicted on an estate map of 1652. The northern section of the gardens, described as the 'wood and walks', contained fishponds, an ornamental building and a collection of statues, all connected by a network of serpentine paths. Hall and gardens were surrounded by a deer park, laid out at the expense of farmland; note the lines of trees, marking earlier field boundaries (Suffolk Record Office).

This said, the increasing popularity of such features from c. 1660 is unquestionable, and by the 1730s some gardens, including some of the most famous in England, consisted almost entirely of ornamental, or at least ornamented, woodland. Gobions in Hertfordshire, designed by Charles Bridgeman from the late 1720s, praised by Horace Walpole and described by George Bickham in his 1750 guide to Stowe as the second

Figure 2.4 Undated map of Gobions in Hertfordshire, showing how the gardens designed by Charles Bridgeman were quite detached from Gobions House itself. They were laid out within an area of existing, semi-natural woodland (Gloucestershire Record Office).

most significant garden in England, is a good example (Figure 2.4).[3] Its layout was originally known from an undated estate map in the Gloucestershire Record Office, engravings and descriptions, as well as from the evidence of the earthworks which still exist within its area, recently surveyed with the able assistance of the Hertfordshire Gardens Trust.[4] But the recent discovery by Anne Rowe of a previously unknown map, possibly Bridgeman's original plan of the site, has cast much new light on its character and development.[5] The Gobions garden, it should be emphasised, was only very loosely connected to the house, by avenues which did not form its major axes; it was an almost free-standing design, one of the reasons, perhaps, why contemporaries rated it so highly. Moreover, the shape and name of the wood in which it was laid out, together with the archaeological evidence, make clear that it was an area of pre-existing ancient woodland, suitably adapted as a garden by having paths and clearings cut into it – a matter to which I shall return shortly.

Vernacular grammars of landscape

When considering any aspect of the history of landscape design, we need always to be awake to the dangers of compartmentalisation, of limiting our historical imagination by concentrating only on a few aspects of human experience. Garden history is not, of course, in itself an enclosed and inward-looking discipline. It enjoys close contacts with other subjects in the humanities (art history, English literature, architectural history) and, perhaps to a lesser extent, in the sciences (horticulture). Nevertheless, there are ways in which our interpretations can be insular. Because many garden historians have a background in subjects which engage principally with high culture, our understandings of particular styles of design, or features in designs, tend to prioritise the influence of classical literature or the bible; of politics and philosophy; or of fashions set in other forms of art and architecture. Up to a point this is fine. But in all periods, although perhaps especially in that with which we are here principally concerned, the later seventeenth and earlier eighteenth centuries, those involved in the design of gardens and landscapes also lived in other, more practical worlds. It was not merely a literary trope of the time that promulgated the pleasures of country living. All members of the landowning elite, from local squires to major political figures with a town house in London as well as extensive family acres in the provinces, took an active interest in rural affairs – in farming and in forestry. It still comes as something of a surprise, but an illuminating one, when we read the letters sent to major political figures in London by their stewards at home, recounting the state of the harvest, the fortunes of lambing, the health of prized livestock, the condition of orchards or the progress of forestry operations.

This larger vernacular world was seldom acknowledged by contemporaries discussing the meaning of gardens and landscape design, by very dint of the fact that it was all pervasive and commonplace, part of the shared experience of the world. We need to make a conscious effort to recapture it in order to understand the full significance and meaning of features like groves and wildernesses. Men approached trees and woodlands with a mass of unarticulated and unconscious mental baggage which might influence their experiences, or their actions, as decisively as anything they had read in Ovid or Virgil, or had seen in a painting by Lorraine. An understanding of wider attitudes to trees and forestry will not in itself 'explain' the popularity of these features in gardens in the post-Restoration period, but it may contribute to our understanding of them, adding a neglected but important dimension. Attitudes to woods and trees, it should also be emphasised, had been developing over several hundred years; they were part of a wider grammar of landscape through which landowners and others understood the physical environment. Such meanings were rooted in practicalities of land management and in long-term inequalities of wealth, power and access to resources; but they were also moderated and shaped by more immediate political changes, economic pressures and environmental imperatives.

We might usefully begin with the observation, at once obvious and of fundamental importance, that wood, including small stuff suitable for fuel, fencing, tools and the like, and timber, trunks and large branches which could be employed structurally, were needed in vast quantities before industrialisation and to an extent even after this.[6] Before the development in the second half of the eighteenth century of a national canal network which allowed coal to be burned as a domestic fuel throughout England, districts lying at a distance from coal fields consumed vast quantities of firewood. Peat, turf and vegetation cut from heaths and commons made a significant contribution, but firewood remained the prime domestic and, in many districts, industrial fuel. Vast quantities of wood were required for other domestic uses, while even in the late seventeenth century, as houses constructed of brick and stone increased in number, prodigious amounts of timber were still required for building, while ever larger amounts were needed to supply Britain's expanding commercial and military fleets. Both before and during the time we are considering the country thus needed huge quantities of wood and timber – could not have existed without it – and nobody walking in a wood, or a grove, would have been unaware of this very basic fact.

Producing wood and timber

Since at least the twelfth century, wood and timber had been produced in a number of ways, the relative importance of which changed over time. The most intensive form of production was represented by woods managed as *coppice-with-standards*, in which the majority of trees and shrubs were *coppices*, repeatedly cut down to a *stool* at or near ground level on a rotation of between eight and 15 years, which rapidly regenerated to produce a crop of straight *poles* (Figures 2.5 and 2.6).[7] Hazel, so useful for fencing and for the wattle-and-daub used in the walls of timber-framed buildings, and ash, excellent for tool handles and firewood, were common components of the understorey, but a wide range of other plants including lime, hornbeam, oak, elm, maple and holly could also be found, depending on local conditions. Woods were, unless very small, felled in sections, to allow constant cropping. Livestock were rigorously excluded from them, or allowed in only at certain times and under close supervision, because they would browse off the regenerating coppices and suppress their regrowth, ultimately destroying them. Standard trees, principally oaks, were also grown in woods of this kind, but they were usually relatively few in number and widely spaced, for their canopy shade would have inhibited the growth of the coppiced understorey beneath. They were seldom allowed to grow to any great age; most were felled when first mature, in the case of oaks, the most common woodland tree, at 70 or 80 years.[8] It is these woods, and more specifically woods of this kind which were established before 1600, which comprise that category of habitat known to ecologists as *ancient woodland*, included

Figure 2.5 Wayland Wood near Watton in Norfolk is still actively managed. The underwood in this portion of the wood has been cut relatively recently (Tom Williamson).

Figure 2.6 Hazel Hurn Wood, Woodrising, Norfolk. Most coppiced woods in England have, like this one, remained un-cut for many decades (Tom Williamson).

within the *Ancient Woodland Inventory*,[9] and thus accorded particular status and protection.

Timber trees were also grown in pasture fields and in particular in hedges, but here they tended to be outnumbered, even in the eighteenth century, by *pollards*. These were, in effect, aerial coppices, cut at a height of around 2 to 3 metres, raised on a trunk or *bolling* in order to keep them out of reach of browsing animals. Oak, ash and elm were the most commonly pollarded trees. The numbers of hedgerow trees in England increased steadily from the fifteenth century as common land, and arable open fields, were progressively enclosed. Thirdly, wood and timber were produced in *wood-pastures*, grazed woodlands in which, once again, the majority of trees were pollarded in order to produce a crop of wood out of reach of browsing livestock. Many wood-pastures, both in the Middle Ages and afterwards, could be found on common land, exploited and managed by local communities.[10] But some were private and of these, the majority were deer parks.

'Park' is a particularly problematic term, in large measure because it changed its meaning over time. Deer parks may have existed in pre-Conquest times, but they were probably a Norman introduction. Either way, by the thirteenth century there were on average one for every three or four parishes in England.[11] They were enclosed areas in which deer were both farmed as a source of venison and hunted for pleasure.[12] They often featured enclosures containing other kinds of prestige foods, pheasants, rabbits, as well as fishponds. Deer were high-status creatures, symbolic of lordship; venison could not be bought and sold on the open market but only received as a gift. Most early parks were well-timbered environments, enclosed with a stout fence, and they usually contained a specialised building called a lodge which served as a base for the park keeper and as a place for the owner to stay while on hunting trips. Some early parks formed a part of the 'landscapes of lordship' laid out around castles and palaces.[13] The principal chambers in some of these residences had windows which were evidently arranged to maximise views across the sylvan, private landscape, and at some places efforts appear to have been made to maximise the apparent extent of the park so viewed. At Ludgershall Castle in Wiltshire, for example, the line of the pale around the northern boundary of the park runs over a small section of skyline, thus emphasising its considerable extent when viewed from the principal buildings;[14] a similar arrangement can be discerned at Castle Rising in Norfolk.[15] Most parks in the twelfth and thirteenth centuries, however, lay in remote places, away from the homes of their owners, largely because they had been enclosed from residual areas of waste beyond the margins of cultivated ground. Even in these locations parks were aesthetically valued and displayed status not only through the deer they contained but also through the size of the boundaries with which they were enclosed – substantial banks, usually with an inner ditch, topped with a fence of close-set split oaks pales.

From the later fourteenth century parks began to be more closely associated with major residences, becoming adjuncts to or even the setting for the mansion and its gardens, and by the middle of the seventeenth century most were to be found next to, or even surrounding, the residences of the nobility or gentry.[16] By this stage they were becoming more ornamental landscapes, but most still contained deer, managed for venison. The landscape parks of the eighteenth century, while they differed in many fundamental ways from the traditional deer park, nevertheless evolved in part from it. In Oliver Rackham's words, eighteenth-century designers were 'heirs to a long tradition' who often adapted existing deer parks and derived key elements of their designs from them.[17] The eighteenth century was simply the 'third age of parks' when 'their design became an art form in the hands of Lancelot "Capability" Brown, Humphry Repton and their contemporaries'.[18]

The long-term aesthetics of woodland

While the long-term importance of private wood-pastures, and their contribution to the designed landscapes of eighteenth-century England, has long been recognised, less attention has been paid in these respects to coppiced woods. But it is arguable that these too had connotations of status, and other features, which ensured that they were integrated into high-status landscapes from an early date. It needs to be emphasised that while wood-pastures might often be found on common land, coppiced woods were almost by definition private property; they were part of the manorial demesne and any common rights exercised over them were circumscribed. This alone suggests that they must have had some significance as symbols of lordship. So far as the evidence goes, most coppiced woods came into existence between the eleventh and the thirteenth centuries as lords enclosed sections of the wider, disappearing tracts of common 'waste' and brought them into more intensive management. The enclosure of woodland was part and parcel of a wider process, of enclosure and allocation of resources, which occurred as settlement and cultivation expanded onto more marginal land and which also included the closer definition of areas of common land and the creation of deer parks. The enclosure of woods, like the enclosure of parks, was thus an act of lordly appropriation, and in this context attention should be drawn to the character of medieval woodland boundaries. Coppiced woods, as I have noted, were enclosed by banks topped by a hedge or fence accompanied by an external ditch, rather than the *internal* one usually associated with park pales. These earthworks are usually interpreted in functional terms; it was imperative that browsing stock were effectively excluded. But the sheer size of many examples is noteworthy, commonly 7 metres or more in width, including external ditch, from which the bank might rise by as much as 2 metres (Figure 2.7). Post-medieval woods are usually bounded by much slighter banks and ditches, little more than a metre or two in width, although they served a similar purpose. These are still topped in some cases with the remains of hawthorn hedges, and in general terms appear little different from the boundaries surrounding contemporary hedged fields. More importantly, if the main purpose of medieval

Figure 2.7 A section of the medieval woodbank at Hockering, Norfolk, marooned within the wood by subsequent expansion. Most coppiced woods were bounded by substantial banks topped by a hedge or fence in order to protect the coppice from browsing stock and to demonstrate lordly ownership (Tom Williamson).

woodbanks was simply to prevent livestock from straying into the wood and consuming the coppice stools, it is unclear why they needed to be significantly larger than contemporary hedges and hedgebanks surrounding private fields, which fulfilled a similar function. We should perhaps consider them not simply as features analogous to field boundaries, but as ones related more to the earthworks raised around other seigniorial possessions in the twelfth and thirteenth centuries, to the boundaries of deer parks and manorial enclosures, or even to the embankments constructed around the smaller castles. Their size may indicate an expectation that livestock would not simply stray from adjoining pastures, but would be deliberately introduced into woods with human assistance. The areas they enclosed had formed part of the common wastes; creating a wood was an act of privatisation and enclosure and a demonstration of lordly power over customary rights of access. Woodbanks may thus have had a symbolic as much as a practical significance, intended to deter trespassers keen to exercise ancestral rights as much as to prevent the entry of animals acting on their own volition.

Of equal interest in a medieval context is the location of coppiced woodland. Woods tended to be rare in upland districts and also to some extent in open-field, 'champion' areas. They were most common in the west of England, especially in the counties bordering Wales, and in the south-east of the country and southern East Anglia, areas of dispersed settlement in which communal agriculture was less developed and pervasive and in which medieval manor houses often stood in isolation, at a distance from other residences, rather than being placed within a village, surrounded by other dwellings. Within such districts, and to a lesser extent elsewhere, woods were generally found on the poorer soils and, in particular, on heavy clays. But they were not usually located towards the *centres* of such areas, but rather towards their margins. In East Anglia, Hertfordshire and Essex, for example, they clustered towards the edges of the valleys dissecting the extensive boulder clay plateau. They were and are, in consequence, more frequent where the plateau was most extensively dissected and correspondingly rare where extensive tracts of level ground occur – here the larger areas of common land tended to cluster.[19] Kenneth Witney has similarly observed how, by late medieval times, most woodland within the Weald of Kent and Sussex survived around the district's periphery rather than, as we might expect, towards its centre.[20] Witney explained this pattern largely in terms of practical economic factors and, in particular, access. The twelfth and thirteenth centuries saw a steady decline in the economic importance of the Wealden woods as swine pastures and grazing grounds and a concomitant increase in the demand for, and thus in the value of, wood and timber. Areas of woodland were enclosed and more intensively managed in places where their products could be transported to markets with relative ease. In the 'central core of the Weald. . . heavy loads were almost undisposable' because of the difficulties involved in moving laden carts in wintertime along difficult clay roads.[21] During the great phase of medieval population expansion in the eleventh, twelfth and thirteenth centuries colonisation was thus directed into the more remote districts, where the 'wastes' were progressively felled and turned to farmland, or else degenerated to open commons; woods survived mainly on the periphery.

Yet in some areas there is a further, complicating factor. In most of the districts in which woodland was abundant, as noted, manorial sites often lay, even at the time of the Conquest, in relatively isolated locations; and the expansion of settlement which occurred between the eleventh and the thirteenth centuries led to the establishment of new sub-manors which might be even more remotely sited, in the wooded peripheries of townships. In a large number of cases, ancient woods and the halls associated with such manors stand in remarkably close proximity, so much so that, in the post-medieval period, a wood has often expanded over an abandoned manorial site, incorporating the earthworks of moats and associated enclosures. One good example is at Hedenham in Norfolk, where the hall was abandoned in the sixteenth century and the wood, which lay immediately to the north, rapidly engulfed its site and that of associated fields and enclosures, evidenced by a map of 1617 showing the final stages in this process.[22] Nearby Gawdy Hall Big Wood in Harleston has expanded over the sites of two manorial sites, both moated, one to the south and one to the north-east. More usually in this county, the manorial hall continues to be occupied, and while sometimes declining over time to the status of tenanted farm it still stands immediately beside an ancient coppiced wood. Some examples: Tindall Hall, located immediately to the east of Tindall Wood (also in Hedenham); Starston Hall, immediately to the south of Starston Wood; or Banyard Hall, beside Banards or Bunwell Wood in Bunwell. Such 'hall-wood' complexes, while they occur in Norfolk, are more common in Suffolk, and especially in Essex and Hertfordshire – districts which were both more manorialised, and more wooded, in medieval times. Rackham has noted this recurrent association of woods and moated sites, suggesting that 'moat makers deliberately chose to live next to woods: doubtless they appreciated the shelter and perhaps the concealment'.[23] But it is unlikely that a wood would have provided an effective hiding place and, given that moats can be found to the north and east as well as to the south of woods, shelter may have been a secondary consideration. The convenience of access which more generally appears to have structured the distribution of woodland in the medieval countryside may at these places have been taken to its logical conclusion, so that coppices were placed almost literally on the manorial doorstep. But in addition woods, as we have noted, were private property, part of the manorial demesne and may thus, like deer parks, have represented sylvan statements of lordly status. It is noteworthy that where the boundary of Gawdy Hall Great Wood passes the southern of the two moats with which it is associated it is raised up higher above the ground outside the wood than anywhere else, accentuating its apparent magnitude when viewed from the moat, perhaps for reasons of display.[24]

The 'great replanting'

Long before the period with which we are here mainly concerned woods, as much as parks, were thus intimately associated with lordship and status; and they may have been valued as aesthetic adornments to a manorial residence, as well as an important economic resource. In the course of the seventeenth and eighteenth century, however, their significance was altered and added to in a number of crucial ways. Historians have long been aware that the period after c. 1660 saw a significant upsurge in tree planting and afforestation, which continued and intensified into the eighteenth century. By the middle years of the century many of these new woods took the form of *plantations*, consisting entirely of timber trees and without a coppiced understorey. Deciduous species, particularly oak, sweet chestnut and beech, were mixed with a rather larger number of conifer 'nurses', usually Scots pine, spruce and larch. In the words of Nathaniel Kent, they comprised

'Great bodies of firs, intermixed with a lesser number of forest trees'.[25] Coppices were still managed and new ones planted, however. Whatever the precise form of this new woodland, the enthusiasm for planting had complex causes. Landowners were fired up by the writings of men like John Evelyn, whose book *Sylva, or a Discourse on Forest Trees* of 1664 was followed (and extensively plagiarised) by a rash of similar texts, including Moses Cook's *The Manner of Raising, Ordering and Improving Forest-Trees* of 1676 and Stephen Switzer's *Ichnographica Rustica* (1718).[26] There was widespread concern that supplies and timber supplies were running dangerously low. Batty Langley in 1728 for example stated that 'our nation will be entirely exhausted of building timber before sixty years are ended'. Men like Philip Miller (1731), James Wheeler (1747), Edmund Wade (1755) and William Hanbury (1758) were also concerned about the military implications of a timber shortage, and throughout the century the government worried about how to provide the vast quantities of timber required by the Royal Navy dockyards.[27] Large-scale planting was thus seen as a patriotic act, and it is noteworthy that the Royal Society for the Encouragement of Arts awarded annual medals for forestry between 1757 and 1835. As Stephen Daniels has argued, moreover, there was a more general association of patriotism and planting in the period after 1660 which went beyond any simple concern for maintaining the nation's naval power, for the planting of trees demonstrated confidence in the future and thus in the new political dispensation brought about by the Restoration of the Monarchy and by the Glorious Revolution of 1688.[28] Planting also expressed confidence in the continuity of ownership on the part of local dynasties; only those who expected to pass on a property to their children and grandchildren would plant over it. Landowners planted to beautify their estates, but also to demonstrate their extent. 'What can be more pleasant than to have the bounds and limits of your property preserved and continued from age to age by the testimony of such living and growing witnesses?', asked John Worlidge in 1669.[29] Such beliefs shaded easily into aesthetics, reinforcing the preference for sylvan scenes going back into the later Middle Ages, which is evidenced by the placing of deer parks beside major residences and, albeit to a lesser extent, of enclosed coppiced woods.[30]

As well as being motivated by a desire for status and to express dynastic continuity eighteenth-century landowners also planted to provide cover for game, especially from the middle decades of the eighteenth century as the pheasant, a bird of woods and woodland-edges, began to take preference over other game as the principal quarry for sportsmen.[31] And as well as all this, we must not forget the obvious point that money could be made from forestry. It is sometimes assumed that landowners made little immediate profit from the new plantations they established, having to wait many decades before the timber matured. This is not entirely true. Because it was difficult to deal with the weeds which competed with the young trees, and also to some extent because significant losses were anticipated from the depredations of rabbits and other animals, the trees in seventeenth- and eighteenth-century plantations were usually planted more closely than would be usual today, sometimes at a density of nearly two trees per square metre. They were then progressively thinned, beginning at ten years, often earlier.[32] The thinned material is often referred to in estate accounts as 'poles', signifying that it was used in a similar way to cuttings from traditional coppices. Nathaniel Kent typically described the great plantation belt around the park at Holkham in Norfolk as comprising 'four hundred and eighty acres of different kinds of plant, two thirds of which are meant to be thinned and cut down for underwood, so as to leave the oak, Spanish chestnut, and beech, only as timber'.[33] In many cases, the final timber crop was itself only thinned, leaving the plantations still dense enough to provide shelter, cover for game or to beautify the countryside. Indeed, estate forestry in eighteenth-century England, as today, was all about balancing a whole range of uses and demands, game management, aesthetics, profit, in such a way that potential conflict was minimised.

With some important exceptions,[34] historians and historical ecologists have emphasised the way in which coppiced woods were eclipsed by the 'new' forestry, based on plantation silviculture, in the course of the eighteenth and nineteenth centuries. Indeed, one of the characteristics of 'ancient' as opposed to recent woodland, according to Natural England, is that it has a coppice and standards structure, as well as having been present in the landscape since before 1600 and containing a range of so-called 'ancient woodland indicators', slow-colonising woodland plants such as wood anemone (*Anemone nemorosa*), yellow archangel (*Lamiastrum galeobdolon*) or primrose (*Primula vulgaris*).[35] In fact, not only were existing coppices generally maintained but new ones were widely established long after 1600 and well into the nineteenth century, partly for practical and economic reasons but also because coppiced woods, within which different areas or sections of the underwood were felled in turn, always boasted a denser structural base than maturing plantations. This could deliver landowners real benefits in terms of game preservation and also in terms of landscape design, creating much stronger visual barriers to frame or curtail a view than the new plantations. Indeed, recent research in eastern England has shown that many registered 'ancient woodlands', perhaps a fifth by number, originated in the eighteenth or nineteenth centuries and that many 'ancient woodland indicators' can colonise these relatively recent coppices, in the right circumstances, with remarkable rapidity.[36]

It is important to emphasise the complex motives for estate planting in the eighteenth century. But it is also important to set these against a broader canvas of social and economic developments which *allowed* landowners to plant on such a large scale. Two in particular are worth highlighting here: changes in patterns of landownership and the spread of enclosure. Although detailed information on the first of these is lacking, there is general agreement among historians that, in the course of the post-medieval period and especially in the period after the Civil War, large landowners tended to increase the size of their holdings at the expense of smaller proprietors, whether minor gentry, or small freeholders and owner-occupiers.[37] This is important as it was, almost by definition, large owners rather than small who undertook significant schemes of afforestation. Only they could afford to put tens or hundreds of acres out of agricultural production, foregoing immediate for medium- or long-term financial benefit. The long-established association between lordship, and woodland, was thus intensified, or at least maintained. But of greater importance was enclosure, of both open fields and commons, which had been continuing in England for centuries but was increasing significantly in scale during the later seventeenth and eighteenth centuries. By 1750 it is unlikely that much more than a fifth of England still remained unenclosed, and thus still open to communal grazing. Enclosure was achieved in a variety of ways but one consequence was that it brought into the hands of large landowners areas of land which could now be put to new uses. In many cases this meant the creation of new areas of woodland, to add to those which already existed on an estate.

Wilderness and grove in context

Readers may be wondering why I have presented this brief and schematic history of woodland and forestry in England in a volume about the use of ornamental woods, groves, wildernesses and the like in garden design. The first reason is simply to emphasise that, while these features clearly carried a range of meanings, the most obvious to contemporaries would probably have been their association with lordship and landownership. The coppiced woods widely established in the period up to the early nineteenth century, and the new plantations which largely but never entirely superseded them, were seldom if ever planted by anyone below the rank of a squire, the lord of a manor. The second is to note that while groves and wildernesses, or features broadly similar to them, featured in Tudor and early Stuart gardens in England, they increased markedly in popularity following the Restoration. The rise coincides, fairly closely, with the more general increase in enthusiasm for forestry expressed in the writing of men like Moses Cook and John Evelyn or Batty Langley, and in the actions of innumerable local landowners. The third reason is to draw attention to the fact that there had long been a close spatial association, where circumstances allowed, between high-status residences and woodland, and even in the eighteenth century this embraced not only woods *within* gardens, but those lying close to them. A number of writers were concerned that owners concentrated on planting the heartlands of their estates, rather than throughout them. Writing of Norfolk, Nathaniel Kent thus observed that while 'gentlemen of fortune' in Norfolk had carried out much tree planting 'in their parks and grounds', the planting of 'pits, angles, and great screens upon the distant parts of their estates, which I conceive to be the greatest object of improvement, has been but little attended to'.[38] Thus the old spatial association of woodland and residence was not simply maintained, but increased, as enclosure allowed the proliferation of woods in open-field areas in which they had often formerly been rare. Lastly, we might note the way in which the shift, in the 1730s and '40s, away from coppicing in groves and wildernesses close to the mansion and towards the creation of 'open groves' without underwood neatly parallels the rise of the new plantation forestry on the wider estate.

In short, areas of planted woodland in and around gardens, and in the wider estate landscape, were closely associated and in practice often difficult to tease apart. Writers like Cook, Evelyn or Langley appear to have made little distinction between different kinds or locations of woods, and the underwood cut from the coppiced woodland in gardens was used as a source of firewood, and for other purposes, in the mansion and on the estate. In this context, it is useful to note again the somewhat blurred line that sometimes existed between anciently established areas of woodland on the one hand, and seventeenth- and eighteenth-century groves, wildernesses and woodland gardens on the other – a blurring exemplified by places like Gobions, where the gardens were entirely cut out of a long-established wood. Because, as we have seen, ancient woods often lay in close proximity to manor houses, many were adapted as ornamental features in the seventeenth and eighteenth centuries, especially in well-wooded counties like Hertfordshire. Cook himself, who worked for much of his career at Cashiobury in that county, laid down directions in 1679 for how rides could be cut through what were clearly pre-existing woods.[39] One of the most famous and best-preserved examples of such re-use in England, albeit much restored in the twentieth century, is at St Paul's Walden Bury; here a pre-existing coppiced wood of c. 12 hectares lying to the north of The Bury contains a pattern of straight hedged *allées* laid out as a *patte d'oie* which framed vistas focused on a variety of statues, garden buildings and the parish church (Figure 2.8). Other Hertfordshire examples include the remarkable garden at Kendals in Aldenham, shown on an undated map of c. 1740, with central vista flanked by serpentine paths connecting small clearings (Figure 2.9). Ground inspection leaves no doubt that this, too, was an ancient wood, lying c. 150 metres to the north-west of the mansion, and still characterised by massive stools of coppiced hornbeam. These places had been medieval manors; and it is thus worth pondering whether the paths cut through the adjacent woods in the seventeenth or early eighteenth centuries really were the first that had connected them with the house. Did medieval owners really only look out at their woods, enjoying the prospect in the distance, without examining them more closely on the ground?

As garden historians we love to compartmentalise developments in design, to force our data into clear chronological patterns. But some of these are in part structured by the character, and the chronology, of the surviving evidence, and linear patterns often turn out to have cyclical aspects; things come in and out of fashion, rather than following a strict sequence of

Figure 2.8 The woodland garden at St Paul's Walden Bury, Hertfordshire, was laid out around 1740 but restored in the early twentieth century. Each of the straight *allées* radiating from the house was focused on a statue or a building, including a temple and the octagonal Organ House within the gardens, and the parish church outside them (Anne Rowe).

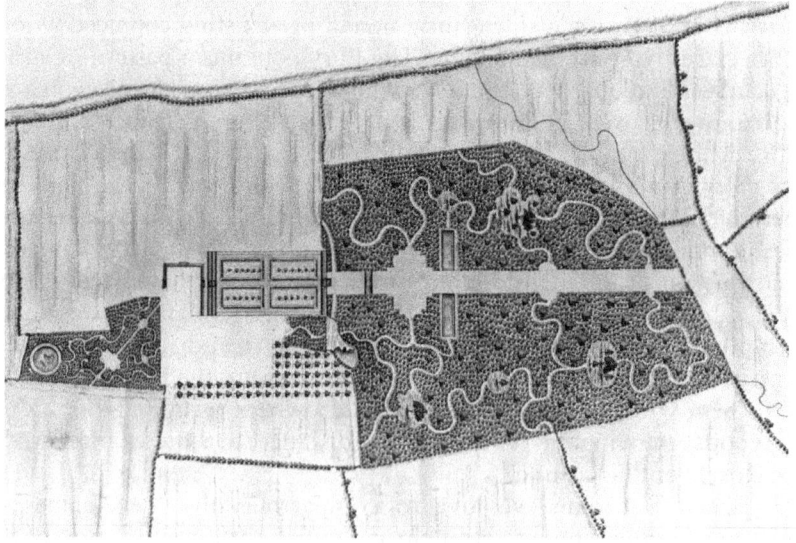

Figure 2.9 Kendals, Aldenham, Hertfordshire: an unfinished plan, probably from the early 1740s. Most of the garden takes the form of an extensive wilderness, created out of an existing area of ancient, semi-natural woodland (Hertfordshire Archives and Local History).

succession. We are accustomed, for example, to discuss how the layout of paths in groves and wildernesses became increasingly serpentine through the 1720s and '30s, a neat and tidy progression. All garden historians should therefore be obliged to examine the map of Somerleyton, surveyed in 1652, which shows the wilderness at the end of the garden – later known as the Image Park, because it was full of statues and sculpture, threaded with decidedly serpentine paths (Figure 2.3).

Conclusion

It may thus be helpful to consider the groves and wildernesses which feature so prominently in seventeenth and early eighteenth-century gardens in longer-term perspective, as part of the more general history of forestry and woodland management in England. Woods had long been closely associated with lordship, and coppiced woods were frequently retained, or established, next to medieval manor houses, at least in districts in which these lay at a distance from other dwellings. This association, and its meaning, continued into the post-medieval period and were probably as significant as any more specific messages conveyed by the design or contents of particular examples within gardens. Woods became more and more important as garden features in the period after 1660, and while there were many reasons for this development it does quite closely mirror the wider enthusiasm for forestry and planting in the period. Because of the proximity of ancient coppiced woods and high-status residences, the former were often converted into woodland gardens or absorbed into the ornamental grounds of the house. But if, as I have argued, they had also been regarded as aesthetically pleasing in earlier periods, then this may simply have been an intensification of existing practice. Recognising continuities as much as changes is crucial for our appreciation of garden history, especially when we attempt to understand the symbolism of gardens; some aspects of woodland were so obvious to contemporaries that they did not need to be mentioned, but we should avoid the trap of failing to see, as it were, the wood for the trees. Neat patterns of chronological development, and an emphasis on change above continuity, in the case of groves as in much else, may prove misleading. Parallels and connections between the worlds of aesthetics and those of the broader economic and social landscape do not in themselves 'explain' the popularity of groves and wildernesses in late seventeenth and early eighteenth-century gardens, but they may *help* to explain it.

Notes

1 David Jacques and Arend Jan van der Horst, *The Gardens of William and Mary* (London: Christopher Helm, 1988), pp. 154–66.
2 Mark Laird, *The Flowering of the Landscape Garden: English Pleasure Grounds 1720–1800* (Philadelphia: University of Pennsylvania Press, 1999), pp. 8–10; John Phibbs, 'Groves and belts', *Garden History* 19:2 (1991), pp. 175–86.

3 Horace Walpole, *History of the Modern Taste in Gardening* (London, 1780), pp. 24–5: George Bickham, *The Beauties of Stow* (London, 1750), pp. 66–7.

4 Peter Willis, *Charles Bridgeman and the English Landscape Garden*, second edn. (London: Zwemmer, 2002), pp. 59–60. 'Map of an Estate belonging to Jeremy Sambrooke Esq.', Gloucester Record Office D1245/FF75.

5 Bodleian Library MS. Maps Herts. a.1. For a full account see Anne Rowe and Tom Williamson, 'New light on Gobions', *Garden History* 40:1 (2012), pp. 82–97.

6 See Brinley Thomas, 'Was There an Energy Crisis in Great Britain in the Seventeenth Century?', *Explorations in Economic History* 23 (1986), pp. 124–52; E. A. Wrigley, *Energy and the English Industrial Revolution* (Cambridge: Cambridge University Press, 2010); Paul Warde, *Energy Consumption in England and Wales 1560–2000* (Naples: CNR-ISS, 2006), pp. 32–9; Paul Warde and Tom Williamson, 'Fuel Supply and Agriculture in Post-Mediaeval England', *Agricultural History Review* 62:1 (2014), pp. 61–82.

7 Oliver Rackham, *Trees and Woodland in the British landscape* (London, 1976); Oliver Rackham, *The History of the Countryside* (London: J.M. Dent, 1986), pp. 62–119.

8 Rackham, *History of the Countryside*, pp. 65–7.

9 George Peterken, *Woodland Conservation and Management* (Cambridge: Chapman and Hall, 1993).

10 For the best account of common wood-pastures, see Patsy Dallas, 'Sustainable Environments: Common Wood Pastures in Norfolk', *Landscape History* 31 (2010), pp. 23–36.

11 Stephen A. Mileson, *Parks in Medieval England* (Oxford: Oxford University Press, 2009); Robert Liddiard (ed.) *The Medieval Deer Park: New Perspectives* (Macclesfield: Windgather, 2007).

12 Jean Birrell, 'Deer and deer farming in medieval England', *Agricultural History Review* 40 (1993), pp. 112–26.

13 Robert Liddiard, *Landscapes of Lordship: Norman Castles and the Countryside in Medieval Norfolk, 1066–1200* (Oxford: Archaeopress, 2000).

14 Paul Everson, Graham Brown and David Stocker, 'The castle earthworks and landscape context', in Peter Ellis (ed.) *Ludgershall Castle, Wiltshire: A Report on the Excavations by Peter Addyman, 1964–1972* (Devizes: Wiltshire Archaeological and Natural History Society, 2000), pp. 97-119.

15 Liddiard, *Landscapes of Lordship*, p. 121.

16 Tom Williamson, *Polite Landscapes: Gardens and Society in Eighteenth-Century England* (Stroud: Alan Sutton, 1995), pp. 22–4. John Fletcher, *Gardens of Earthly Delight: The History of Deer Parks* (Oxford: Windgather, 2011), pp. 162–65.

17 Rackham, *The History of the Countryside*, pp. 126–28.

18 Oliver Rackham, *Woodlands* (London: Collins, 2006), pp. 139–41.

19 Peter Warner, *Greens, Commons and Clayland Colonization* (Leicester: Leicester University Press, 1987), pp. 5–9; Anne Rowe and Tom Williamson, *Hertfordshire: A Landscape History* (Hatfield: University of Hertfordshire Press, 2013), pp. 21–8; T. Williamson, *Environment, Landscape and Society in Early Medieval England* (Woodbridge: Boydell, 2012), pp. 223–30.

20 Kenneth P. Witney, 'The woodland economy of Kent, 1066–1348', *Agricultural History Review* 38 (1998), pp. 20–39.

21 Witney, 'Woodland economy', p. 20.

22 Norfolk Record Office, 1761–61; Oliver Rackham, 'The Ancient Woods of Norfolk', *Transactions of the Norfolk and Norwich Naturalists Society* 27 (1986), 161–7; Gerald Barnes and Tom Williamson, *Rethinking Ancient Woodland: The*

Archaeology and History of Woods in Norfolk (Hatfield: University of Hertfordshire Press, 2015).

23 Rackham, *History of the Countryside*, p. 363.

24 Barnes and Williamson, *Rethinking Ancient Woodland*, pp. 187–89.

25 Nathaniel Kent, *General View of the Agriculture of the County of Norfolk* (London: Crouse, Stevenson and Matchett, 1796), p. 87.

26 Moses Cook, *The Manner of Raising, Ordering and Improving Forest-Trees* (London: Peter Parker, 1676); Stephen Switzer, *Ichnographica Rustica* (London: D. Browne, B. Barker and C. King, W. Mears, et al., 1718), p. 336.

27 Philip Miller, *The Gardener's Dictionary* (London: The Author, 1731); James Wheeler, *The Modern Druid* (London: The Author, 1747); Edward Wade, *A Proposal for Improving and Adorning the Island of Great Britain; for the Maintenance of Our Navy and Shipping* (London: R. and J. Dodsley, 1755); William Hanbury, *An Essay on Planting* (Oxford: S. Parker, 1758).

28 Stephen Daniels, 'The Political Iconography of Woodland in the Eighteenth Century', in *The Iconography of Landscape*, ed. by Dennis Cosgrove and Stephen Daniels (Cambridge: Cambridge University Press, 1988), pp. 51–72.

29 John Worlidge, *Systema Agriculturae* (London, 1669), p. 72.

30 Mileson, *Parks in Medieval England*.

31 Peter Bernard Munsche, *Gentlemen and Poachers: the English game laws 1671–1831* (Cambridge: Cambridge University Press, 1981), p. 8–27.

32 Tom Williamson, *The Archaeology of the Landscape Park* (Oxford: Windgather, 1988), pp. 184–85.

33 Kent, *General View*, p. 90.

34 Especially Ted Collins: see E. J. T. Collins, 'The coppice and underwood trades', in G.E. Mingay (ed.), *The Agrarian History of England and Wales Vol. 6, 1750–1850* (Cambridge: Cambridge University Press, 1989), pp. 484–501; and E. J. T. Collins, 'The wood-fuel economy of eighteenth-century England', in *L'uomo e la Foresta Secc. XIII–XVIII*, ed. by Simonetta Cavaciocchi (Florence: Le Monnier, 1996), pp. 1097–121.

35 Nature Conservancy Council, *Inventory of Ancient Woodland* (Peterborough: Nature Conservancy Council, 1981).

36 Adam Stone and Tom Williamson, 'Pseudo-Ancient Woodland and the *Ancient Woodland Inventory*', *Landscapes* 14:2 (2013), pp. 141–54.

37 John V. Beckett, *The Aristocracy in England 1660–1914* (Oxford: Blackwell, 1986), pp. 43–90; Heather Clemenson, *English Country Houses and Landed Estates* (London: Croom Helm, 1982).

38 Kent, *General View*, p. 87.

39 Cook, *Forest-Trees*, p. 121.

3   The sacredness of groves

David Jacques

The majesty of mature trees *en masse*, that is in groves, is everywhere felt. There can be few readers of this chapter who have remained immune to the sensation, and it is no wonder that sacred groves are revered in many cultures throughout the world.

Our predecessors had some elaborate metaphysical observations in the England of the mid-seventeenth century, as evinced in numerous writings of the time, notably those of John Parkinson (1567–1650), Robert Burton (1577–1640), Sir Thomas Browne (1605–1682) and above all John Evelyn (1620–1706).

Few modern authors have ventured into this territory, an exception being Douglas Chambers, whose research into Evelyn and his contemporaries led him, for example, to Thomas Traherne (1637–1674). He edited a manuscript copy by Traherne in the Bodleian Library for publication as 'Groves' in Toronto in 1987. In *The Planters of the English Landscape Garden*, Chambers expanded upon Evelyn's essay 'An Historical Account of the Sacredness and Use of standing Groves, &c.', drawing attention to his 'physico-theological' beliefs.[1]

'Awfull & sollemne reverence'

Traherne's copy manuscript was of a passage in *A Treatise Containing the Original of Unbelief, Misbelief, or Mispersuasions, Concerning the Attributes of the Deity* by Thomas Jackson (1579–1640), Platonist theologian and president of Corpus Christi College, Oxford in the 1630s. Jackson was wary of the allure of poetry and groves, but was well aware of a letter by Seneca (4 BC–AD 65) 'On the God within us', and its third paragraph:

> If ever you have come upon a grove that is full of ancient trees which have grown to an unusual height, shutting out a view of the sky by a veil of pleached and intertwining branches, then the loftiness of the forest, the seclusion of the spot, and your marvel at the thick unbroken shade in the midst of the open spaces, will prove to you the presence of deity.[2]

Several themes can be discerned here: the ancient wood, the close canopy, the solitary observer and the sense of some divine power preparing the mind for meditation. Evelyn was to assert much the same:

> For our owne part we find it by experience, & professe it that there is nothing strikes a more awfull & sollemne reverence into us, then the gloomy umbrage of some majesticall groves of goodly & tall trees . . . extreamely apt to compose the mind, & infuse into it a kind of naturall Devotion, disposing to prayer, and profound meditation:[3]

Elsewhere:

> We will endeavour to shew how the aire and genious of gardens operat upon humane spirits towards virtue and sanctitie, I meane in a remote, preparatory and instrumental working.[4]

Evelyn scoured the Bible and Classical texts for affirmation of the religious power of groves in antiquity and found many, sufficient for him to be able to assert that: 'the *Patriarchs* themselves did *ab initio* (as 'tis presumed) retire to such places to compose their meditations, and celebrate their sacred mysteries, both of prayer and sacrifices'. Many religions planted groves for the purpose of meditation or worship, a prime example being that of Abraham:

> *Abraham* did but imitate what the children of God had practised before him, when he planted himselfe at the *Quercetum* of *Mambre*: Gen: 13. . . and settled his abode at Barsheba, he design'd a certaine place for Gods divine service: And there the text says. *He planted a Grove and called upon the name of the Lord* etc.[5]

'Pious ecstacies, silent & profound contemplation'

Jackson mentioned the 'solitariness of the place'. Although two or more could discourse within a grove, several authors wrote of the experience as being solitary, a meditation. For example Robert Burton's 'Abstract of Melancholy' was of several verses, all concerning thoughts 'all alone', as

> When I go musing all alone
> Thinking of divers things fore-known.
> When I build castles in the air,
> Void of sorrow and void of fear,
> Pleasing myself with phantasms sweet,
> Methinks the time runs very fleet.
>
> > All my joys to this are folly,
> > Naught so sweet as melancholy.[6]

Figure 3.1 John Evelyn, 'Prospect of Wotton Garden & House towards the East from yᵉ meadow by yᵉ roadside' (1643) (British Library Add MS 78610 C, image c13545–92).

Aged 23, Evelyn had made a 'solitary recess' at his boyhood home at Wotton: 'I made (by my Bro: permission) the stews & receptacles for Fish, and built a little study over a Cascade, to passe my Malencholy houres shaded there with Trees, & silent Enough.'[7] This introduces us to the cult of the melancholy (Figure 3.1).

Melancholy, literally 'dark bile', produced the state of mind that would today be referred to as depression. It could also be understood to induce deep contemplation, seriousness and solemnity. Roy Strong, with his perception as an art curator, drew attention to Marsilio Ficino (1433–1499) whose writings emphasized the latter interpretation and allied it to the character of genius. Those with artistic or intellectual pretentions were anxious to experience, or at least give the impression of, melancholy as an aid to their work and had themselves painted in dark colours and broad-brimmed hats.[8]

Deep contemplation could be induced not only by groves, but also by caves and grottoes. Seneca had written that

> if a cave, made by the deep crumbling of the rocks, holds up a mountain on its arch, a place not built with hands but hollowed out into such spaciousness by natural causes, your soul will be deeply moved by a certain intimation of the existence of God.[9]

Sir Francis Bacon promoted the idea of caves in gardens as an aid to philosophical reflection. His *New Atlantis* (1627) described a fictional country that had reached an almost perfect state. The 'father of Solomon's House' revealed some of the 'preparations and instruments' that laid the foundations of knowledge:[10]

> We have large and deepe *Caves* of several Depths: The deepest are sunke 600 Fathome: And some of them are digged and made under great Hills and Mountaines: . . . for *Prolongation of Life*, in some Hermits that

choose to live ther, well accommodated of all things necessarie, and indeed live very long; By whom also we learne many things.

Bacon's physician, Dr William Harvey (1578–1657), who discovered the circulation of blood, formed caves in his own garden, perhaps in the 1630s:

He had a house heretofore at Combe, in Surrey, a good aire and prospect, where he had Caves made in the Earth in which in Summer time he delighted to meditate.[11]

Harvey was also physician to the Earl of Arundel, who made caves at Albury in about 1635 'wherein he delighted to sit and discourse'. Arundel narrowly escaped death when the soft sand within collapsed. One of Arundel's sons, Charles Howard, must have started on his garden at the Deepdene, Dorking, Surrey, in 1655 or before, for that year Evelyn 'went to Darking, to see M^r. Charles Howard's Amphitheatre Garden . . . he shew'd us divers rare plants; Caves, an Elaboratory'.[12] Evelyn believed in the beneficial effects of caves and grottoes. He started corresponding with Sir Thomas Browne in January 1657/8 in order to obtain his help over certain chapters in his projected *Elysium Britannicum* (which he never took to the printers and had to wait until 2001 for publication). As he remarked to Browne one of his aims was to explain: 'How caves, grotts, mounts, and irregular ornaments of gardens do contribute to contemplative and philosophicall enthusiasme'.[13]

One actual example of such 'enthusiasme' that Evelyn gave was Backbury, in Herefordshire. Although the view one way was over a wide and fertile vale,

At a furlong distance from this sweete and naturall Garden, breakes a most horrid and deepe precipice, fitted for Solitary Grotts and Caverns, and upon the Top of this is a prospect over a most desolate Country, called the Vale of Misery, full of poore and wild Cottages seated on many lesser hills, nemorous and perruked with woods, and other vast objects of rocks, caves, mountaines and stupendous Solitudes fitting to dispose the behoulder to pious Ecstacies, silent & profound contemplation.[14]

This response to a scene of desolation may seem perverse, but similar ones a century later would be explained as the emotion of the 'sublime', and provide a reminder that intervening literary fashions often mask continuities in popular appreciations. Evelyn was not alone in finding worth in a view that might by others be thought to be hideous, as is shown by a brief passage in a work by Florence Estienne Méric Casaubon (1599–1671). Although he was principally concerned with foretelling the future, he stated as if obvious:

That the sight of vast objects, as rocks and mountains, and wild prospects, and the attent consideration of some natural object in a solitary place, doth dispose some men to Ecstasie, that is, transport their thoughts beyond their ordinary limits, and doth raise strange affections in them, I know to be most true.[15]

Arbours and bowers

Lengthy meditation suggested seating of some kind. In his preface, 'To the Reader', Burton described Democritus being discovered in his bower by Hippocrates, the 'father of Western medicine':

> *Hippocrates relates* at large in his Epistle to *Damegetus*, wherein he doth expresse, how coming to visit him one day, he found Democritus in his garden at Abdera, in the Suburbs, under a shadie bower, with a book on his knees, busie at his study, sometimes writing, sometimes walking. The subject of his book was melancholy and madnes.

This scene was chosen for the frontispiece of Burton's book (Figure 3.2). Evelyn recalled Edmund Spenser's lines:

Where in the thickest Covert of that Shade,
There was a pleasant Arbour, not by Art,
But of the Trees owne Inclination made,
Which knitting their ranke Branches part to part.[16]

Evelyn's own 'solitary recess' has been mentioned, and in describing the delights of groves for *Elysium Britannicum* he mentioned 'Sepulchers, Oratories, & whatsoever dos render it sacred & sollomne, & such as may best compose the mind for devotion and profoundest contemplation'.[17]

> Such arbours and bowers were by design simple and as close to Nature as comfort permitted: they should be distinguished from permanent garden structures such as summerhouses, though no doubt there were variations and compromises.

Converse with 'good angells'

There remains the question of the means by which groves or grottoes affected the human mind. Generally the mechanism was assumed to be through the spirit world.

As God (or the gods) held all knowledge, something discovered was a gift from Him, not an original thought by the discoverer. Inspiration was the act of knowledge or insight being passed down. In 1514 Albrecht Dürer portrayed melancholia as the state of waiting for inspiration to strike (Figure 3.3).

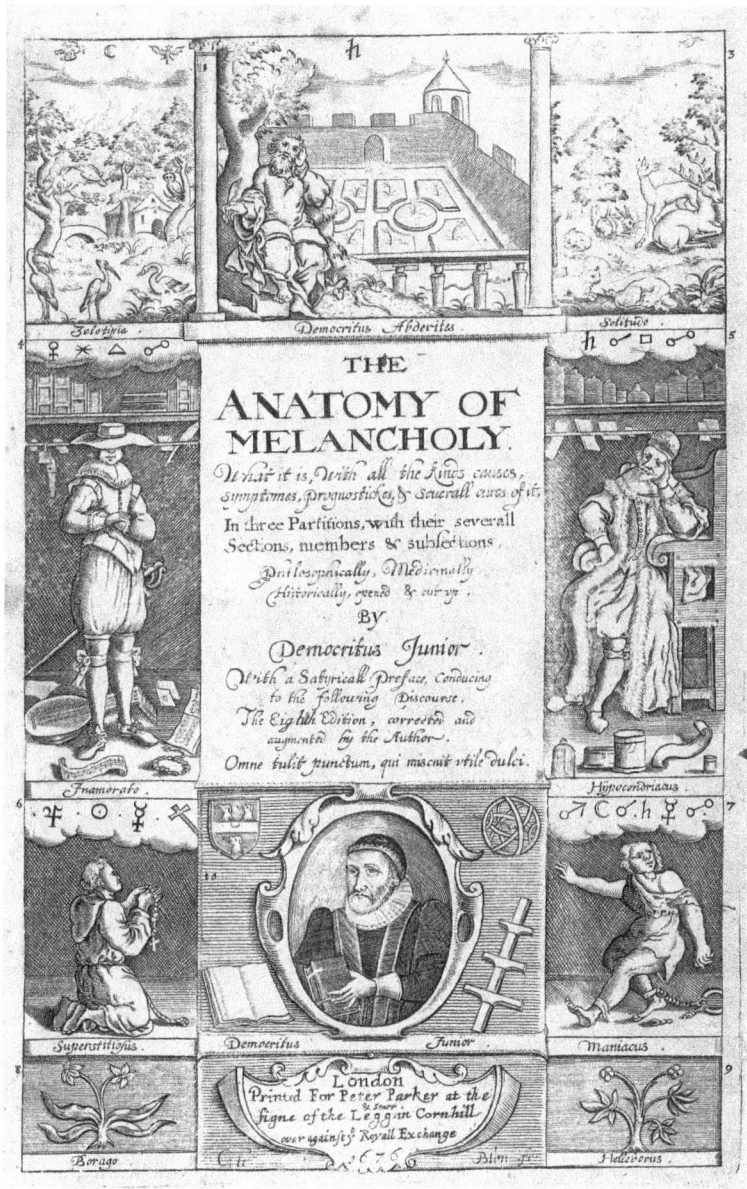

Figure 3.2 Frontispiece of Robert Burton's *The Anatomy of Melancholy* (London, 1638).

Figure 3.3 Albrecht Dürer portrayed 'Melencolia', melancholia, as the state of waiting for inspiration to strike, 1514 (http://commons.wikimedia.org/).

In Greek culture the Muses were seen as passing on inspiration from the
gods. The Romans invoked the concept of an external creative 'genius',
linked to the sacred or the divine. In Judaism and early Christianity creation was the sole province of God; humans were not considered to have the

ability to create something new, just to be the conduit of God's revelation. Most notably, the Bible was written by divine inspiration.

Jehovah communicated knowledge through his angels, as in the frontispiece of John Parkinson's *Paradisus Terrestris* of 1629 illustrates (Figure 3.4).

Figure 3.4 Frontispiece of John Parkinson's *Paradisus Terrestris* (London, 1629).

Burton similarly saw angels (or devils) as the means of transmission, as his 'Digression of the nature of Spirits, bad Angels, or Devils, and how they cause Melancholy' made clear:

> How far the power of spirits and devils doth extend, and whether they can cause this, or any other disease, is a serious question, and worthy to be considered: for the better understanding of which, I will make a brief digression of the nature of spirits.[18]

In his letter to Browne, Evelyn enthused about gardens and groves facilitating intercourse with 'good angells':

> *Elysium, Antrum, Nemus, Paradysus, Hortus, Lucus*, &c., signifie all of them rem sacram et divinam; for these expedients do influence the soule and spirits of man, and prepare them for converse with good angells; besides which, they contribute to the lesse abstracted pleasures, phylosophy naturall and longevitie:[19]

This quality of gardens could extend from religious sentiments to those required for wise rule. Edmund Waller (1606–1687) suggested in his poem on St James's Park that Charles II had gained wisdom to rule from holding Court in the walks in front of the palace:

> In such green palaces the first Kings reign'd,
> Slept in their shades, and Angels entertain'd:
> With such old Counsellors they did advise,
> And by frequenting sacred Groves grew wise;
> Free from th'impediments of light and noise
> Man thus retir'd hid nobler thoughts imploys:[20]

'The trapps & gins of sacrilegious superstition'

The Puritans were ideologically committed to improved husbandry, some being active in improving techniques and others in promoting orchards. Nevertheless, creating wealth in this way was no reason to boast. Their unease at showy display even extended to the use of flowers for decoration on the grounds that they hinted at unnecessary luxury and heathenism, and when used in church on feast days or at funerals their very gorgeousness was a distraction from the word and worship of God.[21] The countervailing view was that they were one of God's fine creations, with usefulness both in physick and in delight. John Parkinson, a committed Protestant, elided the Puritan's suspicion by these arguments in *Paradisus Terrestris* (1629).[22] After all, had not Adam been placed in Paradise, where flowers provided both nourishment and pleasure to the senses?

Jackson, mentioned above, mistrusted the claims for the spiritual virtues of groves. He did so not because he doubted the communion with spirits, but because he feared that the spirits encountered in groves might be 'wicked spirits' that encouraged idolatry and heathendom.

Thus he acknowledged that 'Every unusuall place, or spectacle, whether remarkably beautifull, or ghastly, imprints a touch or apprehension of some latent invisible power'.[23] The nature of such power might well, though, lead to superstition rather than true religion:

> And because superstition can hardly sprout, but from the degenerate, & corrupt seeds of Devotion, wicked spirits did haunt these places most which they perceived fittest for devout devotions.

Indeed, Jackson argued, idolatrous religions were known to adhere to their groves unaware or oblivious of the wickedness of the spirits, and to adorn them:

> As sight of such Groves & fountains, as Seneca describes, would nourish affection: so the affection naturally desirous to enlarge itself would, with the help of these spirits, sleights & Instigations, incite the superstitious to make their groves more retired & sightly . . . And their appearance being most usuall, when mens minds are thus turned to Devotion: the eye would easily seduce the heart to fasten his affection to the places wherein they appeared, as more sacred than any other. And to the spirits thus appearing, as to the sole lords & owners of the delightfull soile, & chief patrons of these bewitching cities, & customes, they thought their best devotions were not too good.

He appealed to the Scriptures to give examples of this misdirection. Groves thus became snares and delusions to be avoided: 'Throughout the stories of the Judges & Kings of Israell, we may observe how Groves were as the banqueting houses of false Gods: the trapps & gins of sacrilegious superstition.'[24]

The Old Testament remedy was to cut down groves and demolish altars:

> And unto Gideon the first to whom this warrant was in particular directed, throw down the Altar of Baal that thy father hath made, & cut down the grove that is by it [Judges 6.v.25]. And Ezekiah, whiles he removed the high places, & brake the Idols, cut down the Groves.
>
> [2 Kings 18.v.4]

Jackson gave a relatively modern example of the benefits of cutting down groves that he had seen in an account of Lithuania when it was still a heathen country. The inhabitants believed that 'set Groves of trees in common woods of unusuall height had such Authoritie, from Antiquitie for their sacred esteem'. It was a sacrilege to cut or burn them, one that would 'instantly provoke vengeance divine'. However one Grand Duke, Jagello (c. 1357–1434), himself baptized, wanted to show that this was mere superstition. He dared to cut down woods and groves, and met with no such retribution. Seeing this, the inhabitants 'grew more tractable, & began to believe the Kings Authoritie & command, becoming at length forward professors of the Christian religion'.

Utility and ideal plans

It is noticeable that metaphysical speculation on groves tailed off after the Restoration. Jackson's wariness does not seem to have impressed the learned of his generation to any degree; rather it was the empirical approach to knowledge, proposed by Sir Francis Bacon, promoted by Samuel Hartlib and leading to the foundation of the Royal Society in 1660.

Evelyn joined the Royal Society and contributed to the enthusiasm. Others knew of his experiments in planting, and he was asked to compose *Sylva* (1664), ostensibly an encouragement to the 'Propagation of Timber in His Majesty's dominions', though providing rather more by way of describing all the trees available for planting at the time, and guidance on their propagation, care and felling. He advised on Moses Cook's highly practical book on *The Manner of Raising, Ordering, and Improving Forest and Fruit-Trees* (1676) which he much respected.

This emphasis on utility was at the very core of the writings of the new science. Another of the authors' traits was the formulation of ideal plans. John Smith, a forestry official, described his in *England's Improvement Reviv'd* (1673). The directions were sufficiently precise for a modern reconstruction to be drawn out.[25] Evelyn himself ventured a plan that he inserted in the 1706 edition of *Sylva*, well known after publication by Miles Hadfield.[26]

Yet Evelyn had not forsaken his convictions on the sacredness of groves; it was simply the other side of the coin to utility. His 1670 edition of *Sylva* expanded the second half of his chapter 'Of Groves, Labyrinths, Daedales, Cabinets, etc.' in *Elysium Britannicum* where he had rehearsed references in Classical literature and the Bible 'to reinforce our esteeme of these goodly shades . . . and historise a little concerning *Groves*, the choicest & most sacred of all the *Hortulan* delights'.[27] As he explained,

> I cannot think to have well acquitted my self of this useful *Subject*, till I shall have in some sort vindicated the honour of *Trees*, and *Woods*, by showing my *Reader* of what Estimation they were of old for their *Divine*, as well as *Civil Uses*.[28]

The result was a new chapter on 'An Historical Account of the Sacredness and Use of standing Groves, &c.', one that he continued to expand in successive editions of *Sylva*. His lengthy researches permitted him to state that the special nature of groves was appreciated by the Romans:

> From hence then began *Temples* to be erected and sought to in such Places; and as there was hardly a *Grove* without its *Temple*, so had every *Temple* almost, a *Grove* belonging to it, where they placed *Idols*, and *Altars*, and *Lights* endow'd with fair Revenues which the devotion of Superstitious persons continually augmented.[29]

Such a conclusion may have inspired the artist of a sketch of 'The antient manner of Temples in Groves', attributed to William Stukeley (Figure 3.5).[30] Bringing the subject home to Britain, Evelyn later expanded on the traditions of the Druids:

> The famous Druids . . . derived their Oak-theology, namely, from that spreading and gloomy-shading tree . . . they chose the Woods and the Groves, not only for all their religious exercises, but their courts of justice, as the whole institution and discipline is recorded by Caesar . . . at every call in a Grove of venerable Oaks, methinks I hear the answer of an hundred old Druids, and the Bards of our inspired ancestors.[31]

Figure 3.5 William Stukeley (attrib.), 'The antient manner of Temples in Groves', c. 1730. Bodleian, Gough Maps 229, f. 322 (David Jacques, reproduced by permission of the Bodleian Library).

Hence, despite being committed intellectually to utility and improvement, Evelyn knew emotionally that groves had some inexplicable power to connect mankind to the divine. 'The Summ of all is, *Paradise* it self was but a kind of *Nemorous Temple* or sacred *Grove*, Planted by *God* himself, and given to *Man, tanquam primo sacerdoti.*'[32]

Notes

1 Douglas Chambers, *The Planters of the English Landscape Garden* (New Haven: Yale University Press, 1993), pp. 45–9.
2 Seneca, Letter 41, in *Moral letters to Lucilius*, translated by Richard Mott Gummere and made available through the Creative Commons Attribution-ShareAlike License.
3 John Evelyn, *Elysium Britannicum, or The Royal Gardens*, ed. John E. Ingram (Philadelphia: University of Pennsylvania Press, 2001), manuscript p. 107.
4 John Evelyn to Sir Thomas Browne, January 1657–58.
5 Evelyn, *Elysium Britannicum*, manuscript p. 104.
6 Robert Burton, 'Abstract of Melancholy', in *The Anatomy of Melancholy*, 1638, before synopsis.
7 John Evelyn, *Diary*, for 23 July 1643.
8 Roy Strong, 'The Elizabethan Malady: Melancholy in Elizabeth and Jacobean portraiture', in *Apollo*. LXXIX, 1964.
9 Seneca, as 2.
10 Francis Bacon, *New Atlantis*, 1627, 31–2.
11 John Aubrey, *Brief Lives*, 1898, under 'William Harvey'.
12 John Evelyn, *Diary*, for 1 August 1655.
13 John Evelyn to Sir Thomas Browne, January 1657–58.
14 Evelyn, *Elysium Britannicum*, manuscript p. 57; see also Peter Goodchild, ' "No Phantasticall Utopia, but a Reall Place": John Evelyn, John Beale and Backbury Hill, Herefordshire', *Garden History* 19:2 (1991), pp. 105–27.
15 Meric Casaubon, *A Treatise concerning Enthusiasme* (London: R. D., 1655), p. 46.
16 Edmund Spenser, *The Faerie Queene*, 1590, Book III. Canto VI.
17 Evelyn, *Elysium Britannicum*, manuscript p. 96.
18 Burton, *Anatomy*, Part I. Sect 2. Sub. 2, p. 39.
19 John Evelyn to Sir Thomas Browne, January 1657–58.
20 Edmund Waller, *A Poem on St. James's Park as lately improv'd by His Majestie*, London: 1661.
21 Jack Goody, *The Culture of Flowers* (Cambridge: Cambridge University Press, 1993), pp. 190, 204.
22 John Parkinson, *Paradisi in Sole, Paradisus Terrestris* (London: 1629), 'To the Courteous Reader'.
23 Thomas Jackson, *A Treatise Containing the Original of Unbelief, Misbelief, or Mispersuasions, Concerning the Attributes of the Deity*, 1625, edition of 1844, Book V, chapter XX, p. 177.
24 Jackson, 178.
25 Peter Goodchild, 'John Smith's Paradise and Theatre of Nature: The Plans', *Garden History* 25:1 (1997), pp. 28–44.
26 Miles Hadfield, *Gardening in Britain* (London: John Murray, 1979), p. 110.
27 Evelyn, *Elysium Britannicum*, manuscript p. 103.
28 John Evelyn, *Sylva, or a Discourse on Forest-Trees* (London: John Martyn, 1679), p. 252.
29 Evelyn, *Sylva*, 1679, p. 259.
30 Bodleian Library, MS Gough 229, f.322 i.
31 Evelyn, *Sylva*, 1679, p. 263.
32 Evelyn, *Sylva*, 1679, p. 256.

4 The history and development of groves in English formal gardens (1600–1750)

Jan Woudstra

It is possible to identify national trends in the development of groves in gardens in England from their inception in the sixteenth century as so-called wildernesses. By looking through the lens of an early eighteenth-century French garden design treatise, we can trace their rise to popularity during the second half of the seventeenth and early eighteenth century to their gradual decline as a garden feature during the second half of the eighteenth century. This chapter shows that their identification as wildernesses at times determined some of their design inspiration, though there were also trends that were adapted from the continent. It records the invention, during the first decade of the eighteenth century, of the use of shrubs in graduated arrangements, positioned according to height, which sparked a trend that came to be known on the continent as the *bosquet à l'angloise*. Later, this was incorporated as one of the prime elements of the pleasure ground of the landscape garden, the shrubbery. A celebration of classical culture in England from the 1710s onwards brought an interest in groves and a new imagery that saw them presented as haunts of dryads (wood nymphs) and satyrs, for which the densely planted continental-type wildernesses were considered to be unsuitable. This review investigates how the changing meaning of groves and wildernesses affected their design and maintenance. It highlights how transnational and local trends interacted with, and bridged, various garden styles.

The theory and practice of gardening (1712)

When Antoine Joseph Dézallier d'Argenville (1680–1765) produced his soon to be famous *La theorie et la pratique du jardinage* in 1709, in which he fleshed out the text according to chapter headings by Jean-Baptiste Alexandre le Blond (1679–1719), he presented a modern world view. It looked at garden making as an art in an aesthetic sense and in its practicalities, and provided clear guidance for the disposition of gardens; but there was no reference to either meaning or symbolism. With an international appeal that led to its being translated into English[1] and German, and with pirated editions in the Netherlands, it promoted gardens, groves and a range of water features in what was described as 'the style of Le Nôtre'. He considered woods and groves as 'the *Relievo* of Gardens', there to contrast with the flatter parts, being pierced with alleys to create 'the Star, the direct Cross, S. *Andrew*'s Cross, and the Goose-Foot', as well as including a range of ornamental features such as cloisters, labyrinths and bowling greens which he increased in a later edition. The woodland areas within these designs were to be neither too small, nor so large as 'to leave great Squares of Wood naked, and without Ornament'. For the treatment of these he provided patterns for six types of woods and groves. These patterns clearly influenced later designs of groves, and by identifying types and trends in the design of groves before the publication of the treatise and afterwards it is possible to investigate innovation in the design of these features and the extent of their modernity.

Six types of groves

Dézallier d'Argenville explained that groves and woods were the 'most noble and agreeable in a Garden', that they offered the 'greatest Relief' in the summer heat and provided the opportunity for walks in the shade. So he saw them mainly serving comfort first and pleasure second. The six types of woods and groves were distinguished with respect to their layout and design. The categories included: 'Forests, or great Woods of high Trees; Coppice-Woods; Groves of a Middle Height, with tall Pallisades; Groves opened in Compartments; Groves planted in Quincunx, or in Squares; and Woods of Evergreens' (Figure 4.1).

'Forests, and great Woods of tall Trees' covered a considerable area of 'at least a League, or many Acres in Compass', densely planted with tall growing species 'which form a very thick tufted Head'. They were generally laid out in a star shape with a large circle in the middle, with ridings for hunting but no hedges or rolled walks. They had a 'wild and rural' character. Normally these woods were sown either broadcast or in lines six feet apart, but the best way was to plant well-rooted plants six feet apart which were left to grow out to 'a lofty Stature' (Figure 4.2).

'Coppice-Woods' were cut down to the base every nine years, a process which was phased by dividing the wood in nine parts so that one could be cut every year. In France there was an obligation to leave sixteen 'Tillers' or individual shoots per acre, plus old standards, which ensured that such woods would gradually be transformed into Forests. These woods were sown or planted in a similar manner to forest woods, but set three feet apart; the tops of the plants were cut back in order to create multiple branches and 'spread themselves to a bushy Tuft'.

'Woods of a middle Height with tall Pallisades' were a type of grove that could commonly be found in (French) gardens. They were 'styled of a Middle Height', so that the selection of lower species and management with judicious pruning enabled them to be maintained at a maximum height of thirty or forty feet. This type of grove was associated with features such

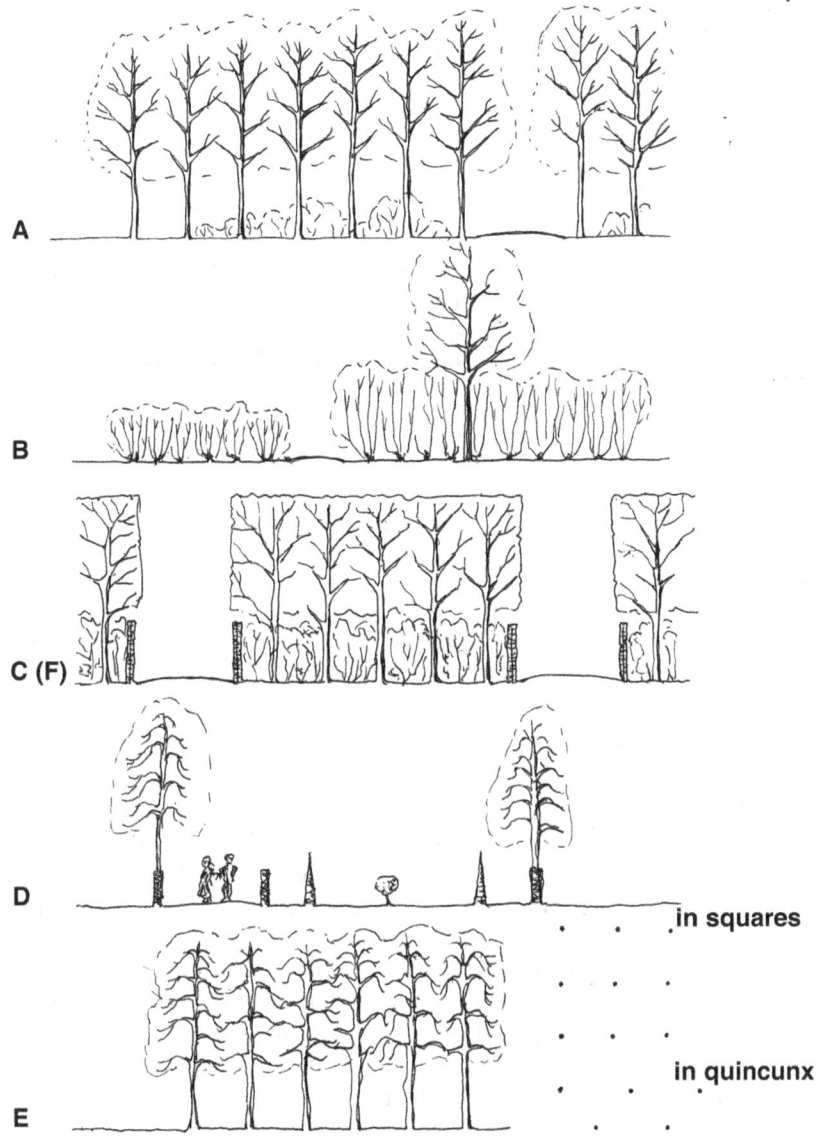

Figure 4.1 Schematic sections of Dézallier d'Argenville's grove types: a. Forests, or great Woods of high Trees; b. Coppice-Woods; c. Groves of a Middle Height, with tall Pallisades; d. Groves opened in Compartments; e. Groves planted in Quincunx, or in Squares; and f. Woods of Evergreens (similar to c, except for the planting) (Jan Woudstra).

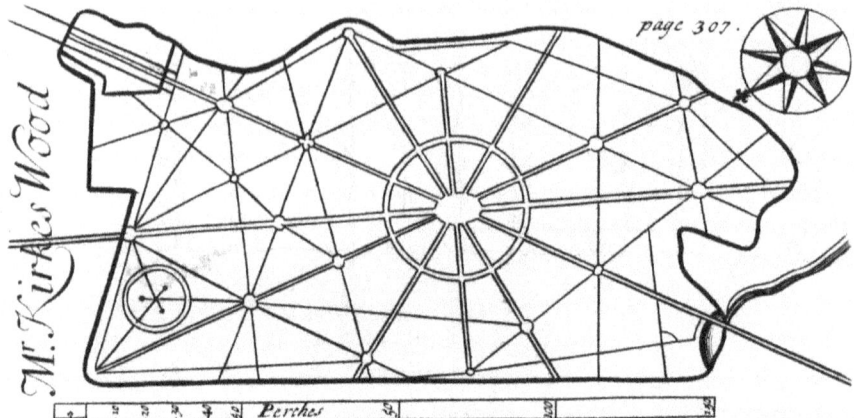

The lines in this Platforme represents the Walkes in M.ʳ Kirke's Wood (call'd Moseley.) neare his House at Cookeridge, (betwixt Leeds and Otley) in Yorkshire. The whole containing about Six Score Akers.

The Double line Walks are about 20 Foot wide, and ij Single lines about 8. Foot wide.

Figure 4.2 Dézallier d'Argenville proposed that 'Forests, and great Woods of tall Trees' were to be laid out in a star shape with a large circle in the middle, with ridings without hedges. John Evelyn depicted such a design, a wood called Mosely at Cookeridge between Leeds and Otley, which however functioned as a coppice wood (John Evelyn, *Sylva* (London, 1729), p. 268).

as 'Halls, Cabinets, Galleries, Fountains', and its squares or quarters were surrounded with hedges and 'Lattice-work' with beautifully finished gravel walks. After carefully laying out and planting these features, the middle of the wood would be planted with elms, chestnuts and so on in rows six feet apart and three feet within the rows. As soon as these were established the areas in between the rows were sown or planted with acorns, chestnuts and others to form a 'Thicket and Brushwood', while the trees within the rows would 'form the Head of the Grove, if Care be taken to trim their Branches, and conduct them to their proper Height' (Figure 4.3).

'Groves that are open, and in Compartments' had trees planted along the walks or alleys surrounding the various quarters, but there were no trees within the squares. The walks were planted with lime trees or horse chestnuts and the hedges were maintained at a height of three or four feet, so that it was possible to see people in other walks. The interior of the squares included compartments and grass cutwork, adorned with pyramid yews and shaped flowering shrubs.

'Groves planted in Quincunce' were those where tall trees were planted in 'several Alleys or Rows' in quincunx formation or at right angles, that is, in

Figure 4.3 Dézallier d'Argenville's 'Woods of a middle Height with tall Pallisades' were not common in England, though they appear to have been the most popular type with gardeners. This example at Chiswick House (John Donowell, *A View of the Three Walks*. (London, 1753)).

squares. There were no hedges and the ground consisted of short turf with an alley through the middle, or of rolled and raked soil. The exactness of the alignment of trees within the rows and between them was 'all the Beauty of them'.

'Woods of Ever-Greens' were considered to be the most desirable, but were not used much because they took too long to establish. They were to be planted 'in the same manner as the others', so no specific detail was provided.

The English scene

Remarkably Dézallier d'Argenville's account relates almost entirely to practical and aesthetic considerations and makes no reference to the meaning or symbolism of these elements in gardens, or suggests that they were places for the imagination. So it presents little 'theory' in a traditional sense. The French word *bosquet* was said in the original text to derive from the Italian *'Bosquetto, a little Wood of small Extent'* and was translated as grove in the English edition by John James.[2] However by that time in Britain the name 'wilderness' was frequently applied to such areas, as it had been since its first use in the sixteenth century. A whole range of other terms were also occasionally used, such as thicket, boscage, coppice, wood or, more rarely, forest, most of which denoted something about the intended character or nature of the area; but it was wilderness that was commonly used and most lasting. The use of this term has confounded generations of observers by the apparent paradox that the groves were included within the confines of the designed landscape, a contradiction heightened by the fact that they were artful, that is, of a highly contrived nature.[3]

One of the earliest surviving accounts of an actual wilderness is by Anthony Watson, c. 1582. Describing the wilderness at Nonsuch, he suggested that it enabled withdrawal 'from those riches of pleasure and prosperity' in the garden 'to less accessible places' suggesting this related to an area outside the garden wall for quiet seclusion and meditation. Working by contrasting opposites he noted that a wilderness might be called a *desertum* although it wasn't deserted, thus clearly referring to the desert in the biblical sense. On the south side there was what probably was a bower, described as 'the canopies' that were 'trimmed in a circular shape'. The walk to the west side appeared as if it had been designed for classical woodland gods and fauns. Birds and other animals were harboured in the many beautiful trees there. There were dwarf apple trees, blackberries and strawberries; there were cherries, oaks, walnuts, ash and elms, periwinkle, pears, hazel, maples, berberis, planes, sycamores, honeysuckle, figs, briars, thorns, dog roses, yew, juniper, elder, box and olive, plums, ferns, vines, Persian fruit and roses. To the north was a large plane tree with wide spreading branches that provided dense shade 'for people to feast under', converse, listen to the birds and animals and see the caged exotic pheasants and partridges. It was also home to a variety of exotic animals, including lions and boars, bears, deer, Indian asses, crocodiles, panthers, wolves, tigers and snakes.[4] In other words Watson's account provided both practical and philosophical reasons for wildernesses in gardens. It clearly worked on the dichotomy of a wilderness in a garden and a garden (of Eden) in the wilderness.

This dichotomy can also be observed in the writings of Francis Bacon, a *homo universalis*, a politician, scientist and philosopher whose essays were much read. In his description of an ideal princely garden in 1625 he proposed that it should be laid out in three sections. There would be 'a *Greene*' at the entrance, the main garden and a third section proposed as a '*Heath or Desart*', which he envisioned as a '*Naturall wildnesse*' without trees, but with thickets of sweet briar, honeysuckle and wild vine. There would be violets, strawberries and primroses planted in the '*Heath*' as these thrive in the shade and are sweet scented. There would be artificial molehills or *Heaps* either planted with wild thyme, or pinks, or germander, or any of periwinkle, violets, strawberries, cowslips, daisies, red roses, lily of the valley, red sweet williams, bear foot, and so on. Some of these heaps were to have small standards planted on top, of roses, juniper, holly, berberis, red currants, gooseberries, rosemary, bay or sweet briar. There would be shaded alleys planted with all sorts of fruit trees all around. Like the Nonsuch wilderness this visionary example reflects a desire for shade and variety despite the absence of trees, but otherwise relies on a similar range of plants and illustrates a longing to include a wide selection of species.[5]

The influence of Bacon's essay can be felt in the writings of Timothy Nourse (d. 1699), who as a Roman Catholic convert was deprived of a fellowship in Oxford and retired to the country. In 'An essay of a country-house' he presented an ideal vision in which he criticized Versailles and celebrated 'the Old *Romans*' and their accomplishments as well as more modern gardens like Frascati and Tivoli. Nourse's ideal garden was, like Bacon's, divided into three sections, with the third being the grove or wilderness that should be 'Natural-Artificial. . . as to deceive us into a belief of a real Wilderness or Thicket, and yet to be furnished with all the Varieties of Nature'. The proposed 'Boscage' was to 'represent a perpetual Spring' with private alleys or walks aligned with beech hedges. There were to be tufts of cypress, 'planted in the Form of a Theater', which probably meant that they were to be graduated according to height. There was to be a fountain in the lower part surrounded by statues. Fir trees were to be distributed 'in some negligent Order', as also laurels, phillyreas, bays, tamarisks, lilac, althea, fruits, pyracantha, yew, juniper, holly and cork oak, together 'with all sorts of Winter Greens' as well as wild vines, laburnum, Spanish ash, horse chestnut, sweet briar, honeysuckle, rose, almond trees, mulberries, and so on. There were also to be 'little Banks or Hillocks' planted with thyme, violets, primroses, cowslips, daffodils, lily of the valley, blue bottles, daisies, as well as 'all kinds of Flowers which grow wild in the Fields and Woods; as also amongst the Shades Strawberries, and up and down the Green-Walks let there be good store of Camomile, Water-Mint, Organy, and the like; for these being trod upon, yield a pleasant Smell'. The surrounding walls were to be planted with ivy, 'Canadensis' [?], phillyreas and the like, that is, a mixture of evergreen climbers and deciduous shrubs.[6] It is clear this wilderness was to appeal to all the senses.

French influences

These examples appear to illustrate an English vision which was partially based on an imagined Italian one, interjected with biblical references. French influence becomes evident in 1625 as a result of Charles I's marriage to Henrietta Maria, daughter of the king of France (1609–1669). She imported various contingents of French artisans, including the garden designer André Mollet (c. 1600–1665), who also worked for Dutch and Swedish nobility.[7] In 1641–1642 he was back in England at Wimbledon Manor, which Charles had acquired for Henrietta Maria, and where he implemented an extensive scheme for the gardens that included a wilderness. This was one of the items described in the Parliamentary Survey of 1649, prior to the sale of royal property during the Commonwealth, and consisted of 'many young trees, woods, and sprays of a good growth and height, cut and formed into several ovals, squares, and angles, very well ordered; in most angular points whereof, as also in the centre of every oval, stands a Lime tree or Elm'. There were eighteen gravel walks in the wilderness and between it and the adjoining maze there was an avenue of lime trees and elms, interspersed with cypress trees, while on the south side it was enclosed with a tall hawthorn hedge.[8]

In his *Le jardin de plaisir* (1651) Mollet elaborated on the detail of *bosquets*, recommending hedges around the quarters of hawthorn, privet, phillyreas and similar, and the interior filled up with shrubs to form '*des boccages*', thickets, in order to attract all kinds of birds so as to create a natural aviary, which he considered much more pleasant than an artificial one.[9] In the English edition of 1670 he added that there were two types of wilderness, one planted with wild trees and the other with all sorts of evergreens. He recommended the evergreens for gardens and the wild trees for parks and more remote places, because these were prone to grow higher and thicker, which he considered unsuitable for a pleasure garden. Arbours within the wilderness would help to provide cool shade during the summer, serving 'either for studious Retirement, or the enjoyment of Society with two or three Friends, a Bottle of Wine and a Collation'. Yet such arbours were not wholly recommended as they inhibited the flow of air and the hawthorn hedge plants did not grow well, looking dead from the inside.[10] It is clear that these wildernesses contributed substantially to the amenities of gardens. As the plants were hardly ever named, their variety seems to have been secondary. The dichotomy of the English notion of wilderness was lost in this example.

This is also the case with various leading practitioners such as John Evelyn (1620–1706), who became one of the leading horticulturalists and foresters of the second half of the seventeenth century. He produced his widely read *Sylva* in 1662, publishing it two years later.[11] It set out to encourage replanting after the general depletion of timber for the navy and other purposes. Evelyn did not use the word wilderness in his writings, but preferred grove instead. As a royalist he had spent the initial years of the Commonwealth on the continent travelling and visiting houses and gardens; after his return to England in 1652 he set out to create his own garden at Sayes Court, Deptford, near London.

Sayes Court included a grove of a modest scale, measuring some 30 by 70 yards, which was laid out roughly in the shape of a double cross, with a circle in the middle around a mount. The width of the main walks was about nine or ten feet, but there were additional narrower walks referred to as close walks and 'Spider Clawes' that led to cabinets with hedges of alaternus. The mount in the centre was planted with bays and surrounded by a laurel hedge; the total of fourteen cabinets each had a great French walnut nearby: twenty-four were planted there. There were over 500 standard trees of oak, ash, elm, service tree, beech and chestnut, amounting to an average planting distance of four to five feet. The walks were lined with trees too, and there were probably hedges, although these were not specified in the description. Additionally there were thickets of birch, hazel, thorn, wild fruits and evergreens.[12] This grove clearly adopted continental practice. While it may have been planted densely in order to anticipate substantial losses, in order for plants to survive in such incredibly dense plantings they must have been kept to manageable proportions by regular pruning.

These close spacings altered the perception of what these groves were all about, so in their translation of a French work on gardening by François Gentil, the horticulturalists and garden designers George London and Henry Wise were able to maintain that *bosquets* and groves were so-called 'from *Bouquet* a Nosegay'. They believed that 'Gard'ners never meant any thing else by giving this Term to this Compartment', describing it as 'a sort of Green Knot, form'd by the Branches and Leaves of Trees that compose it, plac'd in Rows opposite to each other'. They defined grove as a 'Plot of Ground, more or less, as you think fit, enclos'd in Palisades [hedges] of Hornbeam, the Middle of it fill'd with tall Trees, as Elms or the like, the Tops of which make a Tuft or Plume'. At the foot of these elms, which were regularly spaced along the hedges, 'other little wild Trees' were planted that formed a tuft within, resembling a copse. Groves might occur in various shapes and forms but were only proper for 'spacious Gardens, belonging to the Men of the highest Quality' because it was very expensive to keep them up.

London and Wise noted there were other types of groves which consisted 'only of Trees with high Stems, such as Elms planted in right Angles'. Horse chestnuts might also be used for this type of planting, which 'forms a sort of little Forest'. The ground surface below it should be 'kept very smooth, and well roll'd', or it might be turf. Such regular groves were particularly suitable near a palace, while irregular groves of this kind were more appropriate in 'great Parks'.[13] Groves 'made into walks' which, when well executed and maintained, 'invite all that see them to walk in their Shade', were normally planted at right angles, with elms spaced at fifteen feet. Trees with tall stems were most appreciated: 'the Stems Ten Foot long at first, afterward you may raise them to Fifteen or Sixteen, always remembring that the tallest Elms are the finest.' The aim was to have a 'Bush at Top', that is, one that was well spread, so that it provided adequate shade.[14]

The mount at Kensington

While this translation was being published Wise himself had already been experimenting with an alternative manner of planting in the wildernesses to the north of Kensington Palace in 1704,[15] an area that had formerly been quarried for gravel. One old gravel pit had been converted to a terraced orangery for setting out greens with a sunken parterre in the base, and as a contrast, 'on the other Side of it there appears a seeming Mount, made up of Trees rising one higher than another in Proportion as they approach the Center. A Spectator, who has not heard this Account of it, would think this Circular Mount was not only a real one, but that it had been actually scooped out of that hollow Space which I have before mentioned'.[16]

The general principle of arranging plants according to height and colour had long been applied, for example in the planting of borders, but this appears to have been the first time that this strategy was adopted for planting trees. While the significance of this may have escaped Wise, who did not mention it even as a footnote to his book, it was soon observed by other gardeners, and the principle was applied by Thomas Fairchild, for example in his 1722 proposal that London squares should be laid out as wildernesses.[17] But this was later.

France and England

In contrast to France, in England there had been a prevalence of gardens divided into walled enclosures, with wildernesses regularly being contained in separately walled enclosures, or just outside in the park. These brick walls were being utilized for fruit growing which had become fashionable from the early seventeenth century and lasted till the 1720s when they started to be phased out. Evidence of the range of wildernesses and groves in some of the foremost contemporary gardens is provided in the bird's eye views of the seventy country seats depicted in Kip and Knyff's *Britannia Illustrata* (1707) (Figure 4.4). These give some general context for groves in British gardens against which Dézallier d'Argenville's types and designs can be measured. There were fifteen with simple squares; ten irregular or maze-like; ten in star and double cross shape; four in the shape of a cross; two with a St Andrew's cross; as well as seven rectangular, not a shape Dézallier d'Argenville referred to. The views suggest that the largest number of wildernesses (nineteen) were planted with fruit trees (which do not receive a mention in Dézallier d'Argenville's treatise); thirteen as groves in quincunx or open groves; eleven as woods; five as groves open in compartments; and three clipped or shorn that appear to represent groves of middle height.[18] Remarkable in these views is the limited number of groves of middle height, despite the fact that they received more attention in the various treatises than other types, and also the apparent absence of coppice woods. The nature of the engravings leaves little opportunity for distinguishing woods of evergreens. Yet it is possible to conclude that when Dézallier d'Argenville's treatise appeared it was not representative of British practice.

National differences were also observed by Stephen Switzer in 1718, when he judged the translation by John James of *The Theory and Practice* as 'the best that has appeared in this or any other Language', with a good layout and considerable judgement, 'but that being writ in a Country much differing, and very far inferior to this, in respect of the Natural Embellishments of our Gardens, as good Grass, Gravel, &c. makes a great Alteration in point of *Design*. Besides there are some considerable Defects in that way of Gard'ning, as well as in the *Designs* themselves, which I shall take more notice of in due Time and Place'.[19]

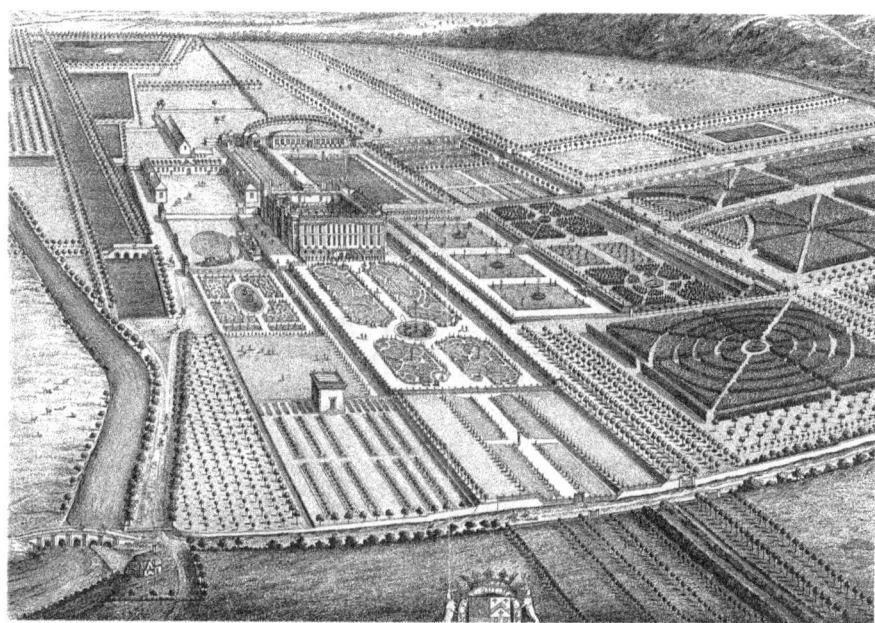

Figure 4.4 Britannia Illustrata by Kip and Knyff showed a good selection of late seventeenth-, early eighteenth-century country seats with wildernesses. This example of Chatsworth shows several going up the slope to the right, each laid out in a star shape with a water feature in the centre; the grove with various circular walks was planted with pine trees only (Johannes Kip and Leonard Knyff, *Britannia Illustrata* (London, 1707), plate 17).

Yet he was inspired in his proposals for 'Rural and Extensive Gardening' by what he referred to as '*La Grand Manier*', the French style of Le Nôtre which consisted of 'large prolated Gardens and Plantations, adorn'd with magnificent Statues and Water-works, full of long extended, shady Walks and Groves', and where 'all the adjacent Country be laid open to View'. He believed that gardens would be more valuable 'if the Beauties of Nature were not corrupted by Art'.[20] So he applauded the fact that the Earl of Carlisle had not followed a design by George London for a star in Ray Wood at Castle Howard, Yorkshire, which according to him would have spoiled it, and instead opted for a 'Labyrinth diverting Model' that respected the natural irregularities of the land and avoided existing trees.[21] This point had also been made by Dézallier d'Argenville, but the engraved plans in the 1712 edition of the book did not reveal this and provided Le Blond's idealized examples,[22] though this was corrected with two examples included in the second edition (Figure 4.5). To Switzer, woods near the house would be 'design'd chiefly for Walking, to be as private as is consistent with its own Nature'; Dézallier d'Argenville saw this primarily as providing shade and

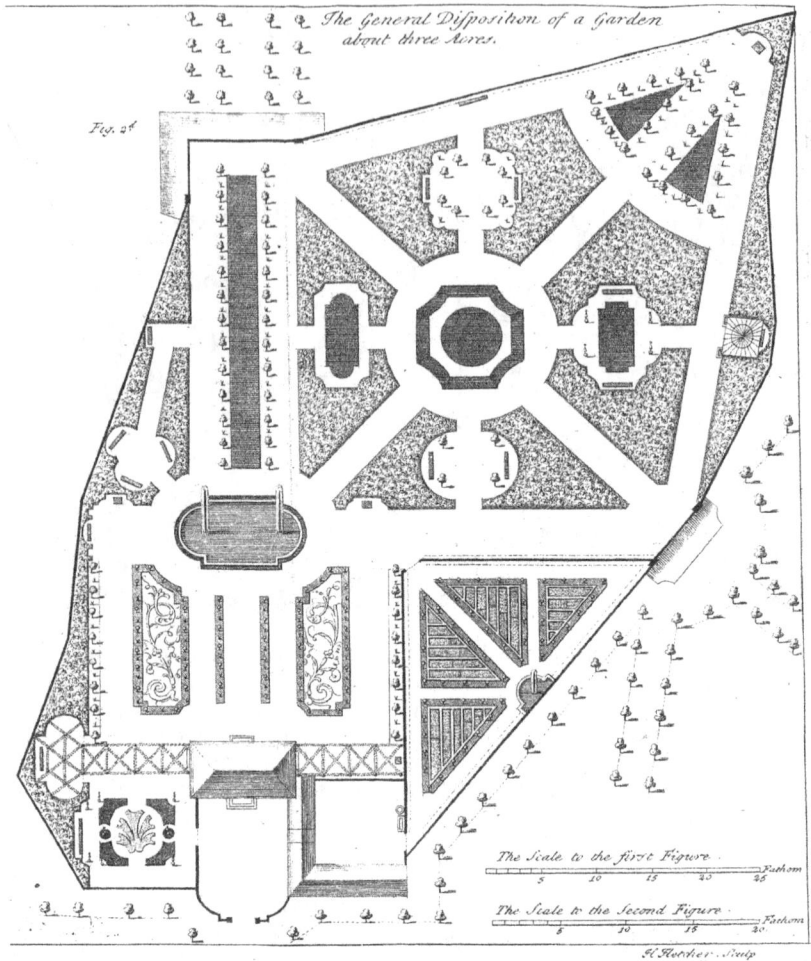

Figure 4.5 Besides a series of idealized examples, Dézallier d'Argenville includes a
couple of irregular layouts in the second English edition of his treatise
(John James, *The Theory and Practice of Gardening* (London, 1728),
plate 5a).

communicating 'a Coolness to the Apartments', not an issue that would
have to be pursued in North Yorkshire.

English woods and rural groves

Stephen Switzer set out to promote 'Rural and Extensive Gardening', and
discussed how to establish woods that were 'more Natural and Rural'.
Rather than the so-called 'Set Wildernesses and Groves', that is, planted

ones, he preferred to have coppices sown, and the 'Witch, *Dutch* Elms and Limes' would be eight, nine or ten feet tall after four or five years, and would out-vie planted ones.[23] Coppices should be planted with oak 'and other natural Furniture of our Coppices' rather than with exotic shrubs. Switzer declared that 'a young Wood springing up 1, 2, 3, or 4 Foot high, is the pleasantest View in Nature, much more pleasant than what it is at full Growth', suggesting that it might incorporate some standards.[24]

Not everyone agreed with this assessment, though, and when Batty Langley reviewed the garden design of the past sixty years in 1728, he concluded that English gardens were *'the worst of any in the World'*. Despite this he claimed that *The Theory and Practice* and Switzer's books were the very best on laying out gardens, while noting that 'even those are far short of *that great Beauty which Gardens ought to consist of'*. He believed this depended on 'the variety of its Parts', which should be disposed in such a way 'as to have a continued Series of *Harmonious Objects*, that will present new and delightful Scenes to our View at every Step we take, which regular Gardens are incapable of doing'. He also noted that traditionally most wildernesses consisted of evergreens, with yews, hollies and other evergreens, as it was these that were grown at the various nurseries, and only rarely forest trees. Other observations were that they were often too far from the house, necessitating passing through 'the *scorching Heat of the Sun'*, and that groves were too regular, like orchards and that instead they should *copy* or *imitate Nature*.[25] Thus several of Langley's pleasure gardens consisted largely of groves with irregular outlines and no three trees in the same line (Figure 4.6).

One design for a rural garden had various open groves of horse chestnuts, of limes, of English elm; other groves had a mixture of standard holly, yew, bay, laurel, evergreen oak, box and phillyrea. All these trees were planted at the base with honeysuckle, sweet briars, white jasmine and various roses, and around the base of the stem were 14–16-inch-wide circles with dwarf stocks, candy tuft, pinks, sweet williams, catch fly, and so on.[26] One plantation with serpentine and straight walks was planted with standards of oak, beech, elm, lime, maple, sycamore, hornbeam, birch, plane and similar, while hedges were planted with English, Dutch and French elms, lime, hornbeam, maple, privet, yew, holly, arbutus, phillyrea and Norway fir. There should also be fruit trees including plums, pears, apples and cherries.[27] He provided a list of scented plants to be planted in groves,[28] as well as a section entitled 'Of the Manner of Disposing and Planting Flowering Shrubs in the proper Parts of a Wilderness' which explained 'the most agreeable and pleasant Manner of disposing and planting of flowering Shrubs'.

In this he adapted the method as established by Wise at Kensington Palace. He divided the flowering shrubs into three classes, of highest growth, middling growth and the low 'Tribe'. The highest plants were to be positioned at the back, far enough from the front to leave room for the other classes. The aim of the planting was to achieve 'a perfect Slope of beautiful Flowers'. Having divided the classes, the next consideration was the colour of the flowers. In order to create the 'greatest Variety' no two plants

Figure 4.6 Several of Langley's pleasure gardens consisted largely of groves with irregular outlines and no three trees in the same line (Batty Langley, *New Principles of Gardening* (London, 1728), plate 3).

of the same colour were to be positioned next to or in front of each other. In the example Langley alternated a white and coloured shrub, providing a sequence of plants that was repeated and thereby formed a rhythmic arrangement. There were low hedges along the walks, with standard trees and jasmine and honeysuckle left 'to run up and about them in a wild and rural Manner'. The inside of the quarters at the back of the shrub planting was to be planted with 'the great Varieties of Forest-trees'.[29] As they were not intended to produce timber and were purely for pleasure they should be planted densely in order to provide shade immediately or form thickets. Elms were to be planted at seven or eight feet, and horse chestnuts at eight, ten, twelve or fifteen feet. In order to produce shade 'the Heads of your Plants' might be cut off so that they would be encouraged 'to spread very much'.[30]

With the gradual development of the landscape garden over the ensuing years the Langley-type groves continued to be adapted, and they ultimately evolved into 'shrubberies' when the term was invented in 1748, after which the name wilderness was gradually phased out. They soon came to be generally adopted and gained further currency, also abroad, through Philip Miller's *The Gardeners Dictionary* (1731). This was particularly so through the various translations of Miller's work in German, Dutch and French; soon after these publications there were references to '*Engelsch Bosch*' (English wood) in the Netherlands, '*englischen Lustgebüsche*' in Germany and '*bosquet à l'angloise* or *à l'anglaise*' in France. There they were ultimately all adapted to the way they were used in the landscape garden in England, and in the English garden abroad,[31] as shrubberies.[32]

While the dense Langley-type planting lost its connection with 'grove', the notion shifted to more open groves, which continued to have supporters, notably Thomas Whately in his *Observations on Modern Gardening* (1770), which was also translated into German and French and was an important guide to the new fashion for the landscape garden. He compared woods and groves and noted that 'the character of a grove is *beauty*; fine trees are lovely objects; a grove is an assemblage of them; in which every individual retains much of its own peculiar elegance; and whatever it loses, is transferred to the superior beauty of the whole'. Differences in shape and colour were only seldom seen as important; they were not to be thinly planted as they would be perceived as a number of single trees, particularly if there was no underwood, but this was not the case with a thick grove. In this instance different shapes and colour might become a consideration, which would also be the case within the groves, as they were also 'delightful as a spot to walk or to sit in'. In order to provide satisfaction only irregular planting would be appropriate.[33] We can see how Whately restricted the definition of groves to a specific type, namely that of the open grove.

This trend had started in the 1710s with a renewed interest in the classics that saw Alexander Pope translate Homer's *Iliad* (1715–1720) and *Odyssey* (1726), in which gardens and groves regained importance as places of the imagination. In an epistle dedicated to Lord Burlington, Pope criticized the densely planted regular (formal) groves: 'Grove nods at grove, each Alley has a brother,/ And half the platform just reflects the other./ The suff'ring eye inverted Nature sees,/ Trees cut to Statues, Statues thick as trees.' The notes in a later edition explain: 'These *groves*, that have no meaning, but very near relation-ship, can express themselves only like twin-ideots by *nods*; which just serve to let us understand, that they know one another, as having been nursed, and brought up by one common parent.'[34] Pope's groves were inhabited by dryads, (wood) nymphs, which provided meaning and imagery that was championed, for example, in the sketches and designs by William Kent.[35] These new poetic groves heralded a break with the French formal grove.

Another example of this was at The Leasowes, where by 1746 the poet William Shenstone had re-created Virgil's grove in a form that remained celebrated into the 1770s. It was described as 'delightful':

> opaque and gloomy, consisting of a small deep valley or dingle, the sides of which are enclosed with regular tufts of hazel and other underwood, and the whole shadowed with lofty trees rising out of the bottom of the dingle, through which a copious stream makes its way through mossy banks, enamelled with primroses, and variety of wild wood flowers.[36]

So rather than considering a grove as a planted feature, it is applied as a setting or garden, with the word grove used as a metaphor, much in the same way as Lady Luxborough used it to describe the whole of his layout.[37]

Conclusion

While its effectiveness as a pattern book and general theory has not been investigated here, this chapter reveals that though Dézallier d'Argenville's treatise is often considered to be both a universal summary and influential, with respects to groves it neither summarized the English situation prior to its publication, nor had a significant impact afterwards. By the time French-style *bosquets*, with very densely planted hedged quarters primarily for walking, shade and coolness, were first introduced in England in the 1640s there was already a characteristic tradition. The English practice of describing groves as wildernesses imbued them with a distinctive notion which influenced their meaning and character and resulted in differentiating design traditions. This unique perspective gave rise to different planting detail and generated some innovative prototypes. Groves of middle height became the prime focus for English horticultural authors who used them as a way to demonstrate their professional expertise in French gardening, but because they were not much depicted in contemporary illustrations, they do not appear to have been commonly applied in England.

Of the other categories or types referred to by Dézallier d'Argenville, groves opened in compartments occurred occasionally, with earlier instances at Ham House and Acklam, for example. Groves planted in quincunx or in squares were probably the favourite type in England, in a well-established practice that pre-dated him. In contrast, and as in France, there were few examples of woods of evergreens, with the best-known example probably that at Castle Howard, familiar through a description from the early 1730s. 'Forests, or great Woods of high trees', were a major focus from the 1660s, and coppices were common practice, but as in France these types were positioned at some distance from the house. Remarkable in Dézallier d'Argenville is the absence of any mention of groves of fruit trees, which appeared commonly in England and the Netherlands.[38]

It was an increased interest in horticultural riches that appears to have encouraged new ways of planting groves with shrubs and trees in a graduated manner, first at Kensington in 1704 and then everywhere else. This became a national tradition that also found its way abroad, first as an alternative way to infill the quarters of *bosquets*, as the *bosquet a l'angloise*, and then in the shrubberies of the English garden. Despite the fact that Dézallier d'Argenville might have heard about this type of planting he did not include this, and it was through various translations of Philip Miller's *Dictionary of Gardening* that these groves were introduced to the continent.

The early eighteenth century's renewed interest in classical culture included a search for meaning in the imagery of the antique, cultivated by Alexander Pope and others. This saw groves as haunts of nymphs and satyrs and required a different arrangement; this coincided with the advent of 'Rural and Extensive Gard'ning' in which gardens featured 'Rural Groves' that were open and planted in an irregular manner, rather than in squares or quincunx, as in the earlier more formal gardens. The new groves, like William Shenstone's Virgil's Grove, were intended as a visual, emotional experience, rather than for ritualistic use, or indeed the practical reasons that had directed design trends previously. By the middle of the eighteenth century the densely planted French-type groves had substantially been replaced by two main types of groves, with one that became known as shrubbery in the English garden, known as English wood or grove on the continent; and the other rural groves that were mainly open, in the English landscape garden. Denser planting could be found there also, but now primarily in clumps and belts.

Notes

1 John James, *The Theory and Practice of Gardening* (London: John James, 1712, 2nd ed. 1728).

2 James, *The Theory and Practice of Gardening* (1712), pp. 48–51; 160–62.

3 For early wildernesses see: Jane Avner, 'Images of the wilderness in some Elizabethan gardens', *Actes des congress de la Société française Shakespeare* 13 (1995), pp. 9–25; Kristina Taylor, 'The earliest wildernesses: their meanings and developments', *Studies in the History of Gardens and Designed Landscapes* 28:2 (2008), pp. 237–51.

4 Martin Biddle, 'The gardens of Nonsuch: sources and dating', *Garden History* 27:1 (1999), pp. 145–83 (174–75).

5 Francis Bacon Lord Verulam, 'Of gardens', in *Essayes of Counsels Civill and Morall* (London: Dent, 1903), pp. 167–76.
6 Timothy Nourse, *Campania Foelix* (London: T. Bennet, 1700), pp. 297–344 (321–22).
7 Laurence Pattacini, 'André Mollet, Royal gardener in St James's Park, London', *Garden History* 26:1 (1998), pp. 3–18.
8 'Parliamentary Surveys, Surrey, No.72. Survey of Wimbledon', in *A History of Gardening in England*, ed. by Alicia Amherst (London: Quaritch, 1896), pp. 315–27 (322).
9 André Mollet, *Le jardin de plaisir* (Stockholm: Henry Kayser, 1651), n.p.
10 Andrew Mollet, *The Garden of Pleasure* (London: John Martyn, 1670), pp. 13–14.
11 John Evelyn, *Sylva, or a Discourse of Forest-Trees, and the Propagation of Timber in His Majesties Dominions* (London: John Martyn and James Allestry, 1664).
12 Evelyn's plan of the gardens (Add.MS 78628 A) can be viewed in detail via the British Library website: www.bl.uk/onlinegallery/onlineex/deptford/p/zoom ify71993.html.
13 George London and Henry Wise, *The Retir'd Gard'ner* (London: Jacob Tonson, 1706), vol. 2 pp. 744–45.
14 London and Wise, *The Retir'd Gard'ner*, vol. 2, pp. 753–54.
15 Stephen Switzer, *Ichnographia Rustica* (London: D. Browne, B. Barker and C. King, W. Mears, et al, 1718), vol. 1, p. 83.
16 Joseph Addison in *The Spectator*, No.477, 6 September 1712.
17 Thomas Fairchild, *The City Gardener* (London: T. Woodward and J. Peele, 1722), p. 12ff.
18 Johannes Kip and Leonard Knyff, *Britannia Illustrata* (London: David Mortier, 1707); see: Jan Woudstra, 'The early eighteenth century wilderness at Stainborough', in *New Arcadian Journal* no. 57/58 (2004–05), pp. 65–86 (70–1).
19 Switzer, *Ichnographia Rustica*, vol. 1, p. vii.
20 Switzer, *Ichnographia Rustica*, vol. 1, pp. xviii–xix.
21 Switzer, *Ichnographia Rustica*, vol. 2, p. 198.
22 James, *The Theory and Practice of Gardening*, 1712 ed., p. 15.
23 Switzer, *Ichnographia Rustica*, vol. 1, p. 272.
24 Switzer, *Ichnographia Rustica*, vol. 3, p. x.
25 Batty Langley, *New Principles of Gardening* (London: A. Bettesworth and J. Batley, 1728), p. iii, xi, x [sic].
26 Langley, *New Principles of Gardening*, p. viii.
27 Langley, *New Principles of Gardening*, p. ix.
28 Langley, *New Principles of Gardening*, p. 184.
29 Langley, *New Principles of Gardening*, p. 181–83.
30 Batty Langley, *A Sure and Easy Method of Improving Estates* (London: Francis Clay and Daniel Browne, 1740), pp. 56, 120–21.
31 Jan Woudstra, 'From *bosquet a l'angloise* to *jardin a l'angloise*; the progression of the mingled manner of planting from its inception to its decline and survival' *Studies in the History of Gardens and Designed Landscapes: An International Quarterly* 33/2 (2013), pp. 71–95.
32 See Mark Laird, *The Flowering of the English Landscape Garden: English Pleasure Grounds 1720–1800* (Philadelphia: University of Pennsylvania Press, 1999).
33 Thomas Whately, *Observations on Modern Gardening* (London: T. Payne, 1770), pp. 46–53.
34 [Alexander Pope], *The Works of Alexander Pope Esq.* (London: J. and P. Knapton, 1751), vol. 3, p. 192.

35 See David Jacques, 'William Kent's "Notion of Gardening": The context, the practice and the posthumous claims', *Garden History* 44:1 (2016), pp. 24–50 (26).

36 Robert Dodsley, *The Poetical Works of William Shenstone* (London: C. Cooke, n.d.), p. xxxi.

37 *Letters Written by the Late Right Honourable Lady Luxborough to William Shenstone, Esq.* (London: J. Dodsley, 1775), see e.g. pp. 56, 57, 278, 284, 346, 349.

38 Jan Woudstra, 'The bosquet and wilderness', in *The Gardens of William and Mary*, ed. by David Jacques and Arend Jan van der Horst (London: Christopher Helm, 1988), pp. 153–66.

5   Colourful groves

The origins of the woodland garden

Brent Elliott

On reflection, one of the most striking things about discussions of tree planting in the gardening literature of the nineteenth century is the virtual disappearance of the word 'grove'. In the first third of the century, one can occasionally come across it, usually in a list, but generally not discussed in detail; thereafter, it disappears even from lists. In the later nineteenth century, you are more likely to find the word 'grove' in discussions of classical gardening, or in works like *The Golden Bough*,[1] which discuss the role of groves in pagan religion, than in a book on modern gardens or even forestry.

Why did 'grove' cease to be a relevant word? It would take a larger and more detailed survey of English usage in the nineteenth century to assign a reason with confidence, but I can hazard a guess. 'Grove', a word hallowed by associations with classical literature, fell out of use as ways of looking at the landscape changed, and a terminology based on visual effect became more widespread; and it also came to be narrowed in meaning to designate a manner of planting that fell out of fashion.

Thus, if we look at the last significant discussion of groves in a nineteenth-century horticultural work, Walter Nicol's *Planter's Kalendar*, whose second and final edition was published in 1820, we find that the word 'grove' has been given an unusually specific meaning: 'a grove is a plantation of trees, whatever be their kind or kinds, which are intended to be trained up with straight tall trunks'. In older usage, 'grove' had been effectively synonymous with 'small wood', but now it appears to have been restricted to a name for a particular method of planting. Nicol's further discussion shows the consequences for landscape design of this restricted meaning:

> This circumstance [the tall straight trunks] will partly determine its extent. If the eye can penetrate through a plantation, it produces a feeling of nakedness. A grove, then, should be of such an extent, or so particularly situated, that, from no side shall the eye be able to penetrate to the other, even were the trees arrived at their full stature, and properly trained. This circumstance shows also the propriety of removing the situation of the grove to a considerable distance from the site of the

mansion-house: It would be no mark of an improved taste to narrow the prospect, by placing a grove in an improper direction.[2]

He went on to virtually conflate 'grove' with 'plantation' and 'mass':

Groves may be constituted of a mixture of trees, like ordinary mixed plantations, but more properly in masses; in which respect, they may be considered as ordinary plantations formed in masses. Indeed, they differ from them hardly in any thing, except that the principals are to be placed rather more closely together. The principals of a deciduous grove should be placed at the distance of six feet; and the interstices filled up with nurses of larch or firs, till the trees in the whole grove be only from three to four feet apart.[3]

The extent to which Nicol's definition of 'grove' reflected a change in general usage, as opposed to being an idiosyncrasy of his own, I cannot say. But if the former, it may provide one reason for the word's disappearance from horticultural literature, as fashions in planting changed.

What happened, of course, was John Claudius Loudon (1783–1843). His proposal of the 'gardenesque' as a new style, formulated in the 1830s, discouraged close planting, whether of flowers or of trees; it shifted emphasis from the mass to the individual specimen, and in this it captured the attention of a generation of gardeners who were already giving specimen trees greater emphasis in planting:

All the trees, shrubs and plants in the gardenesque style are planted and managed in such a way as that each may arrive at perfection and display its beauties to as great advantage as if it were cultivated for that purpose alone; while at the same time, the plants relatively to one another and to the whole scene . . . are either grouped or connected on the same principles of composition as in the picturesque style, or placed regularly and symmetrically as in the geometric style.[4]

There is no reason why this stylistic change need have affected woodlands, but aesthetic systems have a tendency to ooze beyond their original confines. At any rate, in the second quarter of the nineteenth century the progressive tendency was for trees to be planted primarily as specimens, surrounded by enough space that they could develop to maturity without their size and shape being limited by their neighbours; for shrubberies to consist of one or two principal plants, the rest being kept as underplanting; and for trees and shrubs to be placed either in avenues or geometric figures, or in groups rather than indiscriminately scattered.

But Loudon introduced, or at least gave an aesthetic justification for, another tendency: the massing of colours in the flower garden. He attacked the mixture of different colours in the same bed, claiming that 'Variety. . . is not produced by mixture, but by a succession of different things. Every part of a mass, formed on the principle of mixture, is the same in appearance, and the general effect monotonous; but every part of a varied whole differs from every other part, and the general effect is harmonious'.[5] Flower gardens, in the age of Loudon, increasingly made each flower bed a solid mass of a single colour. The same principle, applied to the grouping of shrubs and trees, would yield masses of a single species or visually similar group. There are some indications that there was already a tendency toward this sort of planting, as witnessed by Walter Nicol: 'Groves may be formed of Larches alone. . . . Groves, when composed entirely of Fir, of any of the kinds, have a better effect, when placed in proper situations, than when firs are mixed with other kinds'.[6] The next generation, however, looked back no further than Loudon when assessing the origins of the style. 'Loudon', said Thomas Appleby in 1859, 'lifted up his mighty voice against this absurdity [mixing], and recommended a more scientific mode – that of grouping shrubs'.[7] But such grouping was not, in the 1820s and 1830s, thought of in terms of colour; at this early stage it was the different shapes of trees that received the greatest attention. Look at an arboretum as late as that of Eastnor Castle, replanted after storm damage in the 1860s, and you will see the contrast between rounded, fastigiate, ball-headed and pyramidal trees as the main theme in planting.[8]

The idea that the principles of colour planning being experimented with in the flower garden might be applied to the wider landscape was not expressed openly until the 1860s, but between the 1820s and the 1850s various experiments in landscape colour were carried out. Picturesque theory, in the hands of Uvedale Price, by making the accepted values of landscape painting the standards for landscape planting, allowed an interest in autumn colour, but discouraged an interest in spring colour, which was deemed to be beautiful, and therefore inconsistent with picturesque landscape. The colour effects sought after were, in the words of C.H.J. Smith, those of the blue distance, 'employing dark greens in the foreground, and shading off with lighter colours in the more remote objects', and the use of prominent points of colour 'for occasional breaks in the way of contrast'. 'White poplars and purple beeches', for example, could be used for such purposes, but 'would be intolerable as clumps or masses', and so too the contrast of darker and lighter shades in foliage. Smith allowed a foliage palette of dark green, green, lively green, light green, brownish green, and silvery green.[9] By mid-century, however, picturesque theorists had enlarged their scope and could occasionally hint at a greater range of colour effects. Smith again: 'We do not mean that the chromatic effects of a flower-garden should be by artificial means elaborated in a park or forest; but there is no want of brilliant tints even in the wildness of nature, as the common furze and broom amply testify' (but note that he excluded variegated plants and purple beeches as 'abnormal tints').[10]

James Mangles described the landscape at Highclere:

> the principal feature of attraction . . . is attained by a profusion of clumps, of American bog plants . . . and wherever a stream of lake is at hand, islands are judiciously introduced, and being thickly planted with these American shrubs, present in the summer one gorgeous mass of reflected floration, the splendour of the tints greatly enriched by the tremulous and varied shadows occasioned by the glistening of the waters, and the brilliancy of the carpeted surface above, reminding one of some of Claude Loraine's [sic] glowing sunsets.[11]

Considering that at this time the rhododendrons available consisted of *Rhododendron ponticum* and such American species as *R. catawbiense* (and the reference to American shrubs makes *ponticum* seem less likely), the imagination would have had to work harder to get a Claude Lorrain effect than in later generations.

This initial phase of colour planning in the landscape may be said to have come to a head in 1841, when the *Gardeners' Chronicle*, in its first year of publication, published correspondence on planting for autumn colour. It also carried correspondence, fateful in its long-term consequences, about the ability of *Rhododendron ponticum* to grow in ordinary English soil, unlike the other rhododendrons then available, which as American swamp plants needed peaty conditions. Philip Frost, the head gardener at Dropmore in Buckinghamshire, weighed in to reveal that he had discovered this long before: 'In the woods here we have, by a little attention, thousands of self-sown seedling *Rhododendron ponticum*. . . . When in bloom, nothing can surpass the beauty of Rhododendrons in woods. . . . It is very easy to fill woods with them, by sowing the seed broad-cast'.[12] Now the main reason why *Rhododendron ponticum* was planted so extensively during the next half-century was because it was seen as superior to the traditional cherry laurel for forming coverts for game, but its ornamental value played some part. The earliest description of Bedgebury in the gardening press (1867) described the use of 'masses of Rhododendrons . . . employed to screen the dams and waterfalls' which formed the infrastructure of the artificial lake, but also remarked that 'an island of Rhododendrons of apparently about half an acre in extent, must be very lovely when in bloom'.[13]

Perhaps the most significant use of rhododendrons in the 1840s and 1850s was at Trentham in Staffordshire, where George Fleming, the inventor of the 'ribbon border' and a doyen of the flower garden, began to extend his works into the wider landscape. Thomas Appleby described the results in 1852:

> Mr. Fleming has carried out the planting masses of shrubs of one colour with good effect, especially with the Rhododendron.

Of this charming tribe, so varied in colour, there is, at Trentham, an enormous number, and in particular situations may be seen a large mass of the white varieties – in another, a mass of purple, another of rose, another of scarlet. These, at the part where they come in contact, are judiciously intermixed, so as to soften and blend the two colours together. This attention to planting trees and shrubs, so as to give masses of breadth and colour, is a mark of the onward march of a higher taste in laying out and planting pleasure grounds; and such men as Mr. Fleming, placed in a position to be able to carry such novel views into effect, may be considered as the benefactors of landscape gardening.[14]

Robert Errington, a few years later, agreed that 'Mr. Fleming has, with a high appreciation of landscape gardening, most judiciously converted a hitherto unmeaning extent of ground into a sort of middle distance, – what I must call transition ground. . . occupied by huge masses of intermediate character'.[15] By the end of the decade, Appleby was able to suggest that Fleming's ribbon borders in the flower gardens at Trentham could furnish models for the planning of shrubberies, and concluded, 'We love to see art and symmetry in our arrangement of flowers in a parterre – each bed to have its counterpart, or *fac simile*, in the corresponding bed'.[16] The suggestion that the flower garden could be the model for the wider landscape had almost been uttered.

An interesting controversy of the 1860s, although it was not publicised enough to have a wide influence, indicates the way in which the aesthetic of the flower garden was spreading beyond its confines. Edouard André had won the competition to design Sefton Park in Liverpool, and in his plans he proposed planting dark-leaved oaks, beeches, and limes along the park boundaries: 'Loose-growing and light-coloured trees reserved for centre of Park, and near the water whiter-leaved trees are to be placed. Thus along the drives will be noticed a succession of colours decreasing towards the centre of the Park'. But the adjudicator, Markham Nesfield, took issue with this point and argued:

Dark colours, as reflecting less light, naturally retire of themselves, while light colours come forward. When near at hand, light or cut-leaved foliage is seen to greater advantage against dark foliage, as regards relief of outline and colour; but for the purposes of perspective the distinction is inappreciable as applied to park-planting, as perspective is entirely dependent upon the state of the atmosphere, and is continually varying according to the weather or the hour of the day. This produces what the artist calls effects, but it is quite out of the range of landscape gardening.

Nesfield in effect repudiated the picturesque convention of the blue distance and invoked instead the principles familiar from the flower garden.[17] The stage was set for an advocate for grouping trees by colour to come forward.

William Paul, the great nurseryman and rosarian at Waltham Cross, whose name is commemorated in such roses as 'Paul's Scarlet Climber', published a series of articles in the *Gardeners' Chronicle* in 1864 on 'picto-rial trees', following it with a second series in 1866–7. While some of the articles dealt with differences in visual form (fastigiate, pyramidal, or weep-ing trees), others dealt with differences in foliage colour, and these lists were greatly augmented in the 1866–7 series: light green, dark green, purple, yel-low, and white leaves; trees for autumn colour; trees with white, yellow, or red bark; and smaller lists of flowering trees and trees with ornamental berries.[18] In all this, Paul avoided a proselytising tone, being more concerned to provide lists of what could be used than to provide rules for their usage.

At a meeting of the Royal Horticultural Society the next year, James Bate-man, the creator of the garden at Biddulph Grange, gave a talk adverting to the subject:

> Examples of richly-coloured leaves were then introduced, in order to show the effect which a dark background has in setting off to advan-tage objects of a lighter shade. We have at least three hardy trees which furnish purple of red-coloured leaves, viz. the Purple Nut, Beech, and what deserved to be better known than it is, the Black Maple of Japan. By way of contrast with these deep rich tints, what light ones could be better than those of the Golden Holly, Yew, Ivy, Yellow Bramble, an effective bush, and Honeysuckle. . .? With materials such as these, not tender requiring glass protection, but hardy, beautiful sylvan scenery might be created wich even persons with comparatively limited means might afford to provide.[19]

In 1870 Paul returned to the subject, with a lecture at a horticultural con-gress the RHS convened as part of its Oxford show, and this time he cast off restraint and became an advocate for colour:

> There appears to me a monotony on the face of our English landscapes arising from one uniform and all-pervading colour – green. This monot-ony I would seek to remove by the introduction of trees with purple, white, and yellow leaves. With the same end in view, I would also plant more freely the transitory red, yellow, brown, and purple tints of spring and autumn, supplementing these effects by the introduction of berry-bearing trees – trees with white, red, black, and yellow berries, and trees with white, red, and yellow bark for winter ornament.

He made explicit the analogy with the planning of colour in the flower garden:

> The arrangement of the colours of flowers in the flower garden has of late years been worked out with wonderful skill and effect. What were our flower gardens in this respect thirty years ago? I remember that results predicted then were considered impossible by the many, although they have been accomplished, and more than accomplished,

long ago. Now, as far as I am aware, no one has yet applied the same principles in the arrangement of trees and shrubs with coloured leaves. I have been told that it cannot be done. But after a long study of the question and numerous experiments, I have come to a different conclusion. . . . I believe that here, as in the flower garden, there only needs a beginning, and progress will be rapid and success certain.

. . . Dark bluish green has a good effect when placed in contrast with light yellowish green; white with dark green; reddish purple with light green; reddish purple with yellow; yellow with dark green. And these contrasts by no means exhaust our resources. I merely quote them from among a number of experiments which I have actually tried and found agreeable to my taste.

. . . Of all the errors to be avoided in the association of colours, I would caution the planter against an arrangement that should present a 'spotty' appearance. Broken lines, or irregular shapes of colour, appear to me more desirable in forming plantations or belts than figures with a more than figures with a more easily definable outline.

And Paul concluded hopefully by saying that it was 'a mere question of outlay, and nothing more, whether variety of colour shall or shall not be extended from the garden to the outer pleasure ground and shrubberies, the hills of plantations, the outskirts of woods and forests, and the most distant mountains and plains' (Figure 5.1).[20]

Figure 5.1 Early twentieth-century flowering shrub planting in Broadway, Worcestershire (Brent Elliott).

What was the response to Paul's proposals? Among those in the audience, whose comments were recorded, were William Barron, the head gardener of Elvaston Castle, who thought that recent tree introductions had provided 'every pigment necessary to form the finest landscape', and D. T. Fish, the head gardener of Hardwicke in Suffolk, who 'complained that the ruin of our landscapes had been the mixed system of planting, sufficient attention not having been paid to distinctness of colour. He did not advocate the introduction of so much green into our gardens'.[21]

And as the 1870s gathered pace, so did the enthusiasm for landscape colour. The nurseryman Augustus Mongredien published *Trees & Shrubs for English Plantations* in 1870, its lists of trees including colour. A reviewer in the *Quarterly Review* praised his 'device of grouping', which since Mongredien gave no advice on the subject must reflect experiments that Mongredien was carrying out himself, and thus no doubt helped to alert the wider public to the new aesthetic, while still keeping conifers prominent:

> But though deciduous trees are the backbone of our native timber, where would be the setting of our picture of reds and yellows, where the supplement and finish to our gardens, lawns, and even parks, but for such extraneous races, as pines and firs, cypresses and junipers?[22]

Another nurseryman, Charles Lee, published a proposal for 'Richly-coloured Avenues' in the *Gardeners' Chronicle* for 1875; his avenues exceeded William Barron's by running to twelve or more rows.[23] The illustration of an avenue of golden hollies at Waterers' nursery, which the *Chronicle* had published two years before, was cited as an inspiration.[24] George Abbey, in 1873, called for groups of trees with coloured foliage.[25] D. T. Fish continued his campaign by attacking the monotony of the traditional shrubbery:

> The best thing that could happen to many of them would be their improvement off the face of the landscape. They are huge blocks of green, and nothing more. . . . Our shrubberies, properly observed and interpreted, are our powerful protests against the mixed style of planting. Individual beauty of form, colour, stature, character has been overwhelmed in a nondescript block.

Many, he said, were already trying 'to cut through the dead sea of monotony with the knife. Thousands of shrubs have been cut down in order to make room for specimens'.[26] An example was Rougham Hall, near Bury St Edmunds, where in the 1880s George Paul, the nephew and rival of William Paul, was cutting out vistas and glades through the existing conifer plantation: 'Landscapes can be cut out as well as planted. . . . Vast numbers of fine specimen Oaks, Spruces, Thorns, &c., have thus been evolved out of crowded masses, and set in the new light of attractive landscapes . . . yesterday it was a mass of verdure or other colour, to-morrow the masses may be moulded into groups, with bewitching glades between'. Rhododendrons

Figure 5.2 Rhododendrons in the woodland garden at The High Beeches, Sussex
   (Brent Elliott).

and blocks of purple and golden foliage were specifically mentioned
(Figure 5.2).[27]

The first garden to be written up for its massing of trees and shrubs by
colour was Waddesdon Manor in Buckinghamshire, whose grounds were
laid out and planted in the late 1870s and early 1880s.

> The older and main drives and walks are of great width, and are planted
> on either side with great masses of shrubs, disposed in groups of a sort,
> the form of the groups being triangular, the base and the point coming
> alternately to the front, and as they are now touching each other, and
> generally each plant in a group is fully developed, they will never look
> better than at present. The species and varieties are – Spiraea Thun-
> bergii, Laurustinus . . ., Golden Yew, a capital variety, backed up with
> Viburnum opulus; Thuia sinensis aurea, Acer Negundo variegata, many
> kinds of Berberis, as dulcis, stenophylla, Hookeri, atropurpurea; purple
> Corylus, Copper Beech in several tints of foliage, mahonia aquifolia.
> Lilacs in profusion, and of the best coloured varieties, grow and flower
> beautifully, the Persian being much used for groups. Double Thorns
> of several shades of colour, amongst them Paul's scarlet, standing pre-
> eminent for vivid colour. . . . The effect is heightened by an under-
> growth of closely planted Gynerium. Pyrus, Cerasus, Laburnum,

Rhododendrons, Azaleas, New Zealand Veronicas, Kalmias, and others far too numerous to mention, form those great masses before spoken of as lining the shrubberies on all the main walks. A plant seldom met with in great clumps is Hippophae rhamnoides, but which is at Waddesdon planted in quantity; one great bank of it will in winter exhibit great masses of its heavily-berried branches of orange-scarlet. . . . The bank had an edging of Spanish Furze.[28]

The shrubs that formed these triangles have long since died, and the great bank of sea buckthorn was planted with daffodils in the twentieth century; but scattered through the landscape at Waddesdon there are still clumps and groups of trees that testify to the original planting by colour block.

Among these is a clump of purple beeches, situated just outside the National Trust's portion of the Waddesdon estate. Although it is difficult to get a photograph to display the fact adequately, there are clones visually distinct in colour planted together. (Alan Mitchell, the great surveyor of British trees, hated purple trees and did not collect data on them with the thoroughness he did with other categories. As a result we have no good survey of purple beeches in this country and cannot tell how many of the cultivars whose names can be found at the end of the nineteenth century still survive or how many were juvenile forms which lost their distinctive colouring at maturity.) Copper and purple beeches, augmented from the 1880s by the purple leaved plum, *Prunus cerasifera* 'Pisardii', became mainstays of the late Victorian and Edwardian garden, in avenues and hedges as well as used as specimens.

Even as work was taking place at Waddesdon Manor, in 1878, the City of London acquired Epping Forest in an effort to preserve it from building. In January 1880, William Paul gave a lecture on the Forest to the Royal Society of Arts, presenting his ideas on how it should be treated. He declared that 'never before in the history of England has such an opportunity been afforded of making a grand national recreation ground', and said that 'while we do not want to change the Forest into a landscape garden, or a garden of any kind, we yet want the experience and skill of the landscape gardener to clear away the superabundant rubbish now painfully prominent'. So the Forest was not to become an arboretum, and the existing historic trees were to be the primary focus of its management. Nonetheless, there was plenty of scope for sympathetic alteration: 'Single trees, groups of trees, groves, avenues, thickets, interspersed and relieved with open glades and wide stretches of pasture, are to be grand distinguishing features of the Forest in the future'.

The colours of the trees now existing in the Forest are, for the most part, a uniform green; and however correct it may be to use such as the groundwork of our operations, it is desirable to secure every shade of colour, from grass green and silvery grey to inky black, as well as the varied tints of the unfolding leaves in spring, and the glow of the autumnal

foliage in autumn. By this means the Forest would present an entirely
different aspect at each season of the year – spring, summer, autumn,
and winter; and no mean end would be accomplished if, in planting,
the pleasing natural characteristics of each section could be brought
out in all the fulness of their individuality and beauty. It is desirable
that we should have here more variety in the young leaves in spring;
greater exuberance of foliage, securing coolness and shade in summer; a
richer glow of red, yellow, and russet brown in autumn; and the shelter
and furnished appearance of evergreens in winter. For instance, masses
of the Corstorphine Plane (Acer Pseudo-platanus luteum), the young
leaves of which are of a bright golden yellow, might be planted for
effect in spring; the Douglas Fir (Abies Douglasii) for breadth of light
and shade in summer, and shelter in winter; the Scarlet Oak (Quercus
coccinea) and Norway Maple (Acer platanoides) – the leaves of the former changing to scarlet, and those of the latter to yellow – to vary and
enrich the scenery with their bright and warm autumnal dress. Beech
groves and Oak groves might be formed from trees already existing,
and there might be added Chestnut groves, Birch groves, Lime groves,
Cypress groves, Cedar groves, Wellingtonia groves, and others almost
within [sic = without] limit. These trees should be planted in masses,
and in such manner that they also vary and improve the character of
the forest scenery.

('Grove', in Paul's terminology, was probably a synonym for 'mass', but
it seems to me significant that it should return to horticultural discourse in
a discussion of colour massing.) Among the other trees he proposed planting were purple beech, and purple cultivars of birch and horse chestnut;
whitebeam for white leaves; rowans, Turkey oaks; for autumn colour, liq-
uidambar, Norway and scarlet maples, tulip trees, *Parrotia persica*, ginkgos;
cypresses and cedars; flowering cherries and crab apples; rhododendrons
and azaleas: 'the Rhododendron (R. ponticum) should specially abound'.[29]
J. T. Boscawen, replying, approved Paul's plans in general but recommended
that conifers be excluded: 'I suppose it may be said that those trees grow
in woods in their native countries; no doubt they do, but they do not in
England'.[30] In the event, Alexander McKenzie, who had designed Alexandra,
Finsbury, and Southwark Parks, became the superintendent of Epping Forest, and during the 1880s he followed Paul's suggestion of clearing 'superabundant rubbish', hacking out undergrowth and creating vistas, though he
did not attempt anything like the planting programme Paul had proposed.
    Boscawen's rejection of conifers, and the reason given, indicate that a
new aesthetic was growing up, which would eventually check the onward
march of the colour planners. Grant Allen, back in the 1870s, had attacked
the bright colouring of Victorian painting and especially the prominence
of 'far too much red, purple, orange, and yellow',[31] foreshadowing the
tonal restriction of so much Edwardian painting; and in the closing years
of the century voices could be heard calling for similar tonal restrictions

in planting. Alexander Dean could ask, 'What can be more lovely than a broad expanse of verdure of lively green, over which hang in rich luxuriance trees laden with foliage of darker and diverse shades of green? Not all the cocknified variegation in the world can equal that for beautify and charm';[32] while William Robinson, who had been an enthusiast for colour in the landscape, was praising 'the glory of our mixed native woodland' by the 1920s.[33] Nonetheless, the Edwardian generation continued by and large to appreciate large blocks of colour, or at least to condemn indiscriminate mixtures of trees and shrubs. Take as examples E. T. Cook: 'How good it would be to plant a whole hillside on chalky soil with grand groupings of Yew or Box, or with these intergrouped, and how easy afterwards to run among these groupings of lesser shrubs';[34] or W. J. Bean: 'The value of grouping – that is, the bringing together of several individuals of one kind – is generally not appreciated'.[35]

The long-term legacy of the late nineteenth-century movement for colour was the early twentieth-century woodland garden. The emphasis on rhododendron planting in the wider landscape provided a precedent for the planting of the innumerable new rhododendron species that emerged from China, beginning in the 1890s; the delight in the contrast of flower and leaf colour made the planting of exotic flowering trees amidst native woodland, or more usually amidst mid-Victorian conifer collections, acceptable and exciting. Leonardslee may have been the pioneering estate, Edmund Loder having begun to plant rhododendrons among the conifers in the 1880s. The other leading woodland gardens were begun within a few years of each other: Wakehurst Place with its purchase by G.W.E. Loder in 1903; Bodnant with the construction of the terrace garden in 1905–14; Minterne Manor from 1906, when Lord Digby began to subscribe to plant collectors' expeditions; The High Beeches in the same year; Sheffield Park from Arthur Soames' purchase of the estate in 1909. After the First World War, their ranks were joined by Exbury, begun in 1919 as a rhododendron breeding outstation for Gunnersbury Park. With all of these gardens, the existing literature has been mainly the work of plantsmen whose interest has been the enumeration of species rather than the style of planting; and as the twentieth century advanced, and the drive for massing of colour dwindled, the later doyens of these woodland gardens turned to newer, more mixed forms of planting that have obscured some of their original effect. But one can find occasional statements about the massing of plants by colour. Patrick Synge wrote that 'one of the most effective things about Leonardslee is the large groups in which many of these [rhododendrons] have been planted. Few, who have seen it in full flower, will forget the magnificent bank of R. Thomsonii'.[36] Arthur Hellyer, describing Sheffield Park, referred to 'the brilliance of rhododendrons, which are not arranged haphazard like a patchwork quilt, but are planted in great beds of a kind for skillfully conceived mass effect'.[37] At a later garden like Exbury, Lionel de Rothschild, while dwelling on the individual specimens, nevertheless referred several times to 'banks' or 'masses' of particular cultivars.[38]

Figure 5.3 The Valley Gardens, Windsor Great Park (Brent Elliott).

After the Second World War, two woodland gardens became famous for massed colour effects, which in the 1950s were still at the height of fashion. In the Royal Gardens at Windsor, Eric Savill and Hope Findlay created the 'Kurume Punchbowl', a hillside planted with Kurume azaleas:

> In a vast natural amphitheatre tens of thousands of Kurumes have been grouped in irregular masses of varying sizes and shapes on a scale seldom seen in gardens. . . . Colours were blended and distributed so that one colour would not predominate in any one area. The possibility of clashes was not considered to be a problem, as it is a well-known fact that where one is mixing a wide range of colours, intermediate shades prevent the clashing of colours which would be intolerable if used in twos and threes.

Lanning Roper described the result as 'One of the major horticultural displays of colour in Great Britain' (Figure 5.3).[39] In the same years, Francis Hanger was planting the recently acquired Battleston Hill area of Wisley with a collection of rhododendrons; the result was displayed on one of the first postcards the RHS issued of its garden. In both cases, the style of planting fell out of favour in the second half of the twentieth century; there has been no attempt to keep the Kurume Punchbowl as solidly massed as originally, and when the 1987 storm levelled much of Battleston Hill, the horticultural press breathed a sigh of relief at the thought that something else could now be done with the site (Figure 5.4).

Figure 5.4 The azaleas on Battleston Hill, RHS Garden Wisley, Surrey, in the 1960s (J. E. Downward, Brent Elliott's collection).

The planting of trees and shrubs in masses of single colours was from one point of view a short-lived movement, characterising the last quarter or third of the nineteenth century; but it left behind the varied legacies of the woodland garden and flowering shrub garden, and its energies did not fully dissipate until the last third of the twentieth century. But it also provided the context in which the idea of the grove re-emerged into British horticultural literature – whether that re-emergence had a long-term effect or not.

Notes

1 James George Frazer, *The Golden Bough: A Study in Comparative Religion* (London, 1890), 2 vols.
2 Walter Nicol, *The Planter's Kalendar; or the Nurseryman's and Forester's Guide*, 2nd ed. (Edinburgh: Archibald Constable & Co., 1820), p. 259.
3 Nicol, *The Planter's Kalendar*, p. 260.
4 John Claudius Loudon, 'On laying out and Planting the Lawn, Shrubbery, and Flower-Garden', *Gardener's Magazine*, vol. 19 (1843), pp. 166–67.
5 John Claudius Loudon, 'Difference between Mixture and Variety', *Gardener's Magazine*, vol. 2 (1827), pp. 309–12, esp. 309–10.
6 Nicol, *The Planter's Kalendar*, p. 260.
7 Thomas Appleby, 'A new arrangement of Hardy Shrubs', *Cottage Gardener*, 18 January 1859, pp. 248–49.
8 For Eastnor Castle, see Thomas Baines, 'Eastnor Castle, Ledbury', *Gardeners' Chronicle*, 19 January 1878, pp. 76–8; 26 January 1878, pp. 107–8; 9 February 1878, pp. 170–71.
9 Charles H. J. Smith, *Parks and Pleasure Grounds* (London: Reeve & Co., 1852), pp. 98–102.
10 Smith, *Parks and Pleasure Grounds*, p. 101.
11 James Mangles, *The Floral Calendar* (London: privately printed, 1839), p. 62.
12 Philip Frost, 'Rhododendrons', *Gardeners' Chronicle*, 6 February 1841, p. 85.
13 John Robson, 'Bedgebury', *Journal of Horticulture*, 3 October 1867, pp. 253–55, esp. p. 254.
14 Thomas Appleby, 'Jottings by the Way – Trentham', *Cottage Gardener*, 11 August 1852, pp. 364–65.
15 Robert Errington, 'Visit to Trentham Hall', *Cottage Gardener*, 4 September 1855, pp. 465–66.
16 Thomas Appleby, 'A New Arrangement of Hardy Shrubs', p. 249.
17 André's plan and Nesfield's comments are held in the Liverpool Record Office. See also Brent Elliott, *Victorian Gardens* (London: Batsford, 1986), pp. 170–71, 180–82.
18 William Paul's first series of articles on pictorial trees was published in the *Gardeners' Chronicle* as follows: 13 August 1864, p. 770 [introductory notice of forthcoming series]; 27 August 1864, p. 819; 3 September 1864, pp. 844–45; 17 September 1864, p. 893; 8 October 1864, pp. 963–64; 29 October 1864, p. 1035; 26 November 1864, pp. 1131–32. Paul's second series, 'Hardy Pictorial trees', appeared as follows: 1 December 1866, pp. 1139–40 (pyramidal and weeping trees); 2 February 1867, p. 103 (light and dark green leaves); 9 March 1867, p. 237 (purple, golden leaves); 20 April 1867, pp. 404–5 (white leaves); 15 June 1867, p. 628 (flowering evergreens); 3 August 1867, pp. 805–6 (flowering deciduous trees); 7 September 1867, p. 926 (climbers); 19 October 1867, pp. 1072–73 (berried trees). The articles were reprinted in Paul's collected papers: William Paul, *Contributions to Horticultural Literature* (Waltham Cross: William Paul & Son, 1892), pp. 219–79.

19 James Bateman, quoted in *Gardeners' Chronicle*, 1 July 1865, pp. 605–6.
20 William Paul, 'On colour in the tree scenery of our gardens, parks, and landscapes', *Journal of Horticulture*, 4 August 1870, pp. 82–3.
21 Paul, 'On colour in the tree scenery of our gardens, parks, and landscapes': audience comments added, pp. 83–4.
22 James Davies, 'Ornamental and useful tree-planting', *Quarterly Review* 142 (1876), pp. 50–83, esp. 66–7 and 70.
23 Charles Lee, 'Richly-coloured avenues', *Gardeners' Chronicle*, 5 June 1875, pp. 716–17.
24 Anon., 'The Knap Hill Nursery', *Gardeners' Chronicle,* 6 December 1873, pp. 1634–35.
25 George Abbey, 'The Purple Beech', *Journal of Horticulture*, 18 September 1873, p. 212.
26 David Taylor Fish, 'On the Massing of Shrubs', *Gardeners' Chronicle*, 31 January 1874, p. 143.
27 David Taylor Fish, 'Rougham Hall, Suffolk', *Gardeners' Chronicle*, 24 October 1896, pp. 491–92.
28 Anon., 'Waddesdon', *Gardeners' Chronicle*, 27 June 1885, pp. 820–22, esp. p. 821; and see more generally, Brent Elliott, *Waddesdon Manor: The Garden* (Princes Risborough: National Trust, 1994).
29 William Paul, 'The Future of Epping Forest', *Journal of Horticulture*, 5 February 1880, pp. 96–8.
30 John Townsend Boscawen, 'The Future of Epping Forest', *Journal of Horticulture*, 4 March 1880, p. 182.
31 Grant Allen, 'Colour in painting', *Cornhill Magazine*, vol. 38 (1878), pp. 476–93, esp. p. 477.
32 Alexander Dean, 'The Purple Beech', *Gardeners' Chronicle*, 17 December 1898, p. 445.
33 William Robinson, 'Exotic trees at Gravetye Manor', *Garden*, 11 June 1921, pp. 290–91.
34 Ernest Thomas Cook, *Trees and Shrubs for English Gardens* (London: Country Life, 1902), p. 11.
35 William Jackson Bean, *Trees and Shrubs Hardy in the British Isles* (London: Macmillan, 1914), I, p. 46.
36 Patrick M. Synge, 'Camellias and rhododendrons at Leonardslee', *Rhododendron Yearbook 1955*, pp. 7–16, esp. p. 12.
37 Arthur G. L. Hellyer, 'A garden on the grand scale', *Country Life*, vol. 109 (1951), pp. 1552–57.
38 Lionel de Rothschild, 'Features of my garden – the home wood at Exbury', *Journal of the Royal Horticultural Society*, vol. 65 (1940), pp. 111–14.
39 Lanning Roper, *The Gardens in the Royal Park at Windsor* (London: Chatto & Windus, 1959), pp. 57–9.

6   Groves as metaphor for the fragmented redwood forests of California

Camilla Allen

A CALIFORNIA song!
A prophecy and indirection – a thought impalpable, to breathe, as air;
A chorus of dryads, fading, departing – or hamadryads departing;
A murmuring, fateful, giant voice, out of the earth and sky,
Voice of a mighty dying tree in the Redwood forest dense.
Farewell, my brethren,
Farewell, O earth and sky – farewell, ye neighboring waters;
My time has ended, my term has come.[1]

This chapter seeks an understanding of groves in the context of North American forests, and specifically through the contrasting stories of *Sequoiadendron giganteum* and *Sequoia sempervirens* groves that were identified in California in the nineteenth century. Two species, leftovers from another epoch: the former, otherwise known as giant sequoias, a collection of tall, barrelled, craggy grizzlies, nestled in the High Sierra Mountains; and the latter, a mist-shrouded expanse of forest made up of dense and slim trunks, the coast redwood (Figure 6.1). From the mid-nineteenth century onwards, the *Sequoiadendron* were rapidly discovered, exploited, celebrated and preserved. Whilst, at the same time, and with much less fanfare, the coast redwood forest was reduced to a fraction of its dominion before efforts were eventually stepped up to ensure its survival.

The concept of the formal and sacred grove played an important part in this story. Grove is a word rich with association and meaning with its roots in the old English for copse or small wood, yet transcending that designation with its association with sacred spaces in both real and mythical antiquity; places of transcendence more powerful than an ordinary woodland or forest. Grove, with its mixed applications combining spatial typologies as well as its cultural and literary weight, traversed the Atlantic Ocean with the conquest and settlement of North America and found new use when describing the flora of that continent. The word is now bound up with the perception and preservation of these two mammoth tree species, relics of an ancient forest and ecological period whose recent history relating to their discovery, destruction, celebration and preservation is illuminated by tracing the designation of them as groves.

Figure 6.1 'the tall trees rise in columns of perfect symmetry and form.' Frontispiece from Richard St. Barbe Baker's 1960 edition of *The Redwoods*. A photograph of coast redwood trees originally taken by the Moulin Studio, San Francisco, no date (Baker Papers, University of Saskatchewan).

It is the fate of the remnants of the coast redwood forest that forms the focus of the latter part of this essay and is used to explore how Western ideas and ideals of groves as monuments, temples and places of ritual, have played a part in actions and tensions relating to wilderness preservation, national parks and private property in California. These ideas are then framed in the context of two very different fragments of the redwood forest: the Bohemian Grove and the Grove of Understanding, exploring how their designation as places of community, culture and ritual has shaped their identity, use and fate and how these ideas relate to the understanding of the wider forest ecology.

Two typologies: spatial and spiritual

The word grove travelled with the explorers and settlers from Europe, eventually finding its way to America where its use is much the same as in England: a place with a high canopy of closely planted trees, normally of one type. This essay makes the case that there are two main typologies and associations with groves, spatial, as described here, but also mystical and intangible, and through unpicking their application in the forests in North America, a specific reading of the way these associations have been applied can be understood as having a specific impact on the landscape of California.

The qualities of a grove are something potentially transferrable to all the forests of the world, with sacred associations taking different forms in different cultures.[2] They are elucidated in Seneca's description of them as a space encountered and recognised as having an inherent meaning and value: 'If you come upon a grove of old trees that have shot up above the sight of the sky by the gloom of their matted boughs, you feel there is a spirit in the place, so lofty is the wood, so lone the spot, so wondrous the thick unbroken shade.'[3] Pagan ritual and classical antiquity became intertwined with Christianity as the 'sacred groves of Europe's barbarian prehistory', in turn enriched the emerging idea of the 'cathedral forest'.[4] Robert Pogue Harrison, in his exploration of trees and forests in the cultural imagination, fuses the forest with the architecture of northern Europe where:

> The Gothic cathedral visibly reproduces the ancient scenes of worship in its lofty interior, which rises vertically towards the sky and then curves into a vault from all sides, like so many tree crowns converging into a canopy overhead. Like breaks in the foliage, windows let in light from beyond the enclosure.[5]

Groves were recognised in the pre-European settlement landscape of North America, described in accounts like that of the seventeenth-century French explorer Chevalier de la Salle as peppering the edge of the forests and prairies, akin to the Garden of Eden.[6] These nature experiences shaped perceptions and created a touchstone from which ongoing human relationships with nature and wilderness can be understood. In the context of this chapter groves are used as a metaphor for understanding both conceptions of forest ecologies and as the inspiration for the preservation of America's presettlement forest, much as Paul Kelsch does in his paper exploring the relationship between landscape architecture and the same North American forests.[7] And, beyond that, the Californian redwood groves can also be seen as being emblematic of ideas of a collective natural global commons, worthy of veneration and respect. But they did not start as such . . .

Giant sequoia groves: a new nation's natural monuments

Hushed reverence was not the initial response to the discovery of the grizzlies in the Sierra Nevada in the early 1850s (Figure 6.2). The miners and their hangers on, eager to reap the rich mineral wealth of the mountains of California, were struck most by the spectacle of the gigantic trees that they stumbled upon in their explorations, and at first treated them more like fairground

Figure 6.2 Untitled. Coloured lantern slide of a Giant Sequoia from the Baker Papers and likely to have been used to illustrate Baker's lectures on the Californian redwoods, no date (Baker Papers, University of Saskatchewan).

curiosities than with any hushed awe. Simon Schama captured the heady mix of exploration and extraction which typified the gold rush experience as stands of the almost grotesquely proportioned *Sequoiadendron giganteum* were discovered in the territory also occupied by the mines of Mariposa. It was the size of the individual trees which claimed the first casualties with an ex-miner, George Gale, 'who saw gold in wood, rather than water or rock'[8] and proceeded to strip a tree known as the Mother of the Forest of its bark to a height of 160 feet, to be transported East and displayed to as a botanical wonder, much to the incredulity of the audiences of the time.[9]

The earliest grove of big trees to be recognised quickly became a tourist attraction, 'the commerce of novelty, not the cult of antiquity',[10] with one stump used as a dance floor and another's prodigious length fashioned into a bowling alley. These diversions, along with the trees themselves, generated new scenic tourism to Calaveras Grove where the meshing of this extraordinary natural heritage started to be intertwined with the emerging ideas and ideals of America's nationhood. The arrival of Thomas Starr King and the discovery of another grove closer to the Mariposa mine meant a second chance; instead of the profane amusements of Calaveras, 'America's own natural temple' had been discovered and would be revered as such. 'Pilgrims', Paul Ketch writes, 'came to venerate the ancient trees, swept up in the fact that they dated back to the time of Christ. The trees legitimised American civilisation, grounding it at a time earlier than the great monuments of Europe'.[11] Writing some years later the Scottish American wilderness explorer and writer, John Muir, described the skinning of the Mother of the Forest in Calaveras Grove as being akin to an environmental Calvary:

> as sensible a scheme as skinning our great men would be to prove their greatness. This grand tree is of course dead, a ghastly disfigured ruin, but it still stands erect and holds forth its majestic arms as if alive and saying, 'Forgive them; they know not what they do'.[12]

Muir also recognised the distinct spatial character and distribution of the giant sequoia as standing 'more or less apart in groves, or in small, irregular groups, enabling one to find a way nearly everywhere, along sunny colonnades and through openings that have a smooth, park-like surface, strewn with brown needles and burs',[13] as well as putting forward the theory that their scattered placement on the mountain where they occupy very fertile, sheltered areas, as being as the result of glacial erosion which broke up the larger community and left a scattering of small groups of trees across the slopes of the Sierra mountains.[14] But this objective naturalism was contrasted with the romanticism with which human history was read in their rings: 'The pious notion that the Big Trees were somehow contemporaries of Christ became a standard refrain in their hymns of praise. . . as if contemporaneity banished geographical distance'[15] linking Judea with California, to the extent that they were seen as being the 'Hebrew tree'.[16]

In 1864 Abraham Lincoln signed congressional legislation to pass over Yosemite Valley and Mariposa Grove to public ownership and ensure free access to people of the entire nation. Frederick Law Olmsted experienced the Sierras whilst working for the gold mining Mariposa Company whilst Central Park was in a state of hiatus.[17] In his role as lobbyist, he supported the establishment of the reserve and commented in his report on the newly inaugurated Mariposa Grove and Yosemite Valley that the benefit to society of 'the occasional contemplation of natural scenes of an impressive character. . . is favorable to the health and vigor of their intellect beyond any other conditions which can be offered them'.[18]

Sequoiadendron giganteum remains the emblem of the early wilderness movement, with John Muir establishing the Sierra Club in 1892 as a means of making the Sierra mountains accessible to those interested in exploring the forests and mountains found within the national park. An early success was the establishment in 1893 of the Sierra Forest Reserve, which covered around four million hectares and included the majority of the giant sequoia groves in the area, with many of the major figures being immortalised in a natural pantheon, including the luminaries 'Thomas Starr King' and 'General Sherman'.[19]

A fractured forest

Economics also paid a large part in the monumentalising of the big trees as, conversely, although they gave up extraordinary volumes of wood the timber's brittleness made it highly unsuitable for lumber (Figure 6.3).[20] On the other hand, the coast redwoods, *Sequoia sempervirens*, were already being plundered at an extraordinary rate, their slimmer trunks yielding much of the material for the expanding metropolises of the West Coast. The discovery of the grizzlies quickly attained the status of legend, with the encounter between Augustus T. Dowd and the first specimen being recorded in James Hutchings's *Scenes of Wonder and Curiosity in California*.[21] By contrast, the early history of the coast redwood was considered of much less merit. Drawn into Schama's account of the grizzlies the presence of the coast redwood appears in the history of the botanical classification of what was then *Sequoia gigantea*, a nomenclature which took in both its extraordinary size as well as memorialising the man who created an alphabet for the Cherokee, Sequoyah; a footnote, rather than a key player.

The ribbon of *Sequoia sempervirens* forest was spread along a length of Californian coast from San Francisco to the Oregon border, an area typified by a cool maritime climate with substantial sea mists that carry miles inland. The name, meaning ever-living, derives from the species' ability to reproduce asexually from root stock and buds, the close stands forming concentric circles as the trees regenerate around themselves. By the end of the nineteenth century, however, the scale of loss started to become evident, and the energy that had amassed behind the big trees was something that John Muir thought should be directed to preserve the coast redwoods:

> The danger these Sequoias have been in will do good far beyond the boundaries of the Calaveras Grove, in saving other groves and forests, and quickening interest in forest affairs in general. While the iron of public sentiment is hot let us strike hard. In particular, a reservation or national park of the only other species of Sequoia, the sempervirens, or redwood, hardly less wonderful than the gigantea, should be quickly secured. It will have to be acquired by gift or purchase, for the Government has sold every section of the entire redwood belt from the Oregon boundary to below Santa Cruz.[22]

Figure 6.3 'Wood that is more durable than stone. The lower tree must have fallen
at least 2,000 years ago.' Lantern slide reproduced in 1960 edition of
The Redwoods (p. 43) demonstrating the unusual qualities of redwood
timber, no date (Baker Papers, University of Saskatchewan).

Song of the Redwood-Tree, by the poet Walt Whitman, encompasses
much of the optimism which was felt when settlers encountered the great
natural wonder and resource that are the redwood forests of California and
which had left them ripe for exploitation. The poem gives a voice to the forest, in which it abdicates its rule and passes it to the men who, with 'crack-
ling blows of axes, sounding musically, driven by strong arms' were taking
ownership of the rich land and transforming it into a place of trade, industry

and agriculture, free from the vices and bloody history of other continents.[23] Initially thought impervious to fire, the coast redwood provided the timber with which San Francisco, Oakland and Berkeley were built, as well as sleepers for the railway and stakes for the vineyards of California, but the fires that destroyed San Francisco proved that it was only the bark which is resistant to flame.[24] The first part of the forest to face near total-decimation was around San Francisco, with few areas left unlogged south or east of the city. At this point it is the fate of the remaining coast redwoods in the north of California that best demonstrates the split in the conception of them as forest or grove: either as still being an integral part of a wider sylvan ecosystem, or perceived as individual groves within a landscape increasingly fractured by the growth of human settlements, agriculture and industry (Figure 6.4).

The Sempervirens Club, established in 1900, was the first preservation group to check the rapid destruction of the coast redwoods in the Santa Cruz Mountains and the first significant victory to preserve the trees was theirs. This was followed by the efforts of one man, William Kent, to preserve the

Figure 6.4 Untitled. It was the construction of a road up the coast towards Oregon which made the remaining coast redwoods accessible to loggers in the early twentieth century, no date (Baker Papers, University of Saskatchewan).

redwoods in a canyon in Marin County. This tree-filled gorge close to San Francisco had survived due to its inaccessibility and was in the ownership of a utilities company until 1908 when Kent, the local congressman, bought the 247 hectares of land and gave it to the state so as to prevent the valley from being damned and used as a reservoir. A condition of the bequest was that it would be named in honour of John Muir, with President Theodore Roosevelt issuing a decree to make it a national monument and to preserve it in its natural state as a recreational public space. The designation of the Muir Woods National Monument as a wood, rather than a grove or forest, marks a point before momentum gathered behind the campaign to save the coast redwoods. Still reflective of the patchwork that the coastal forest was becoming, the area was of a significant scale and no one species is singled out in its name and instead it correlates more accurately with Muir Woods' role in the region's watershed and ecosystem. Another area saved by a combination of private donations matched with state funding was the Rockefeller Forest in Humboldt Redwoods State Park, which was created using two donations from John D. Rockefeller Jr and covers 10,000 acres.[25]

Coast redwood groves

Consecration played a significant part in the recognition and designation of groves from 1917 onwards with the formation of the Save the Redwoods League. Susan Schrepfer's history of the campaign to save the redwoods starts with the journey made by Madison Grant, Henry Fairfield Osborn and John Campbell Merriam from San Francisco in search of a forest unparalleled in creation, the journey made possible by a new highway which was snaking through the hills, allowing for exploration into a relatively unsettled area of the state. The men emerged into the valley of Bull Creek, an area with redwood trees so dense that shade was turned into twilight and which, with the sound of logging nearby, they quickly determined to save.[26]

Schrepfer's account sets out in detail the hard-won and ongoing fight that coalesced in 1919 with the formation of the Save the Redwoods League, an organisation which brought together the burgeoning national parks movement, the growing American middle class and the emerging tensions between private and public property in California. Conserving or preserving the redwoods in such a way that balanced public and private economic interests was a pressing concern of the League,[27] and groves played an immediate part. It was set out that the first officers of the League, including Steve Mather of the National Parks, would make rescuing 'outstanding redwood groves' the priority of the organisation.[28] The focus on groves, rather than larger swathes of forest, could be demonstrative of the lack of knowledge at the time of the redwood forest's ecology, but most likely also reflected the initial scale of success that the League could aspire to, as well as being the most evocative symbol of the forest.

This focus on a very intimate scale is demonstrative of the agency with which the League operated. There were longstanding tensions between the wilderness preservation movement, spearheaded by John Muir, and the more economic conservation model favoured by Gifford Pinchot of the U.S. Forest Service. A major contributor to this tension between preservation and conservation was the use of state funds to purchase forest lands which left the taxpayers accountable if the acquisition proved to be uneconomic.[29] Groves, in this instance, provided a scale on which the group could fund-raise and campaign which operated almost in parallel with the machinations on a state level to establish park preserves containing the redwoods.

The first grove to be established by the Save the Redwoods League was Bolling Grove in 1921, named in honour of the first officer of rank to fall in the First World War, and this quickly set the precedent for the establishment for the creation of places such as Richardson Grove and the Garden Club of America Grove to name just some of the one thousand dedicated groves in the publically accessible redwood parks of California. The significance of the act of dedication is evident in the speech made by Richard Nixon in 1969 when presenting the Lady Bird Johnson Grove to the nation, urging those gathered there 'to stand here in this grove of redwoods, to realize what a few moments of solitude in this magnificent place can mean, what it can mean to a man who is President, what it can mean to any man or any woman who needs time to get away from whatever may be the burdens of all of our tasks, and then that renewal that comes from it'.[30] The act of nominating groves became an entrenched part of the Save the Redwoods League's operation, with the average annual purchase and dedication of groves between two and ten acres. This is a key way for the League to raise funds to purchase wider swaths of the ecosystems and watersheds that the trees sit within, as well as funding education and research projects.[31] This act, in which groups and individuals can be immortalised, now plays a critical role in sustaining the coastal climate upon which the redwood trees depend.

Temples and places of ritual and imagination: the Bohemian Grove and the Grove of Understanding

It is a rather prosaic interpretation of groves, mostly in line with the spatial designation and secular memorialising that is manifest in the fundraising strategies of the League. But this rather straightforward reading, without its associations with gods or ritual, is not the only manifestation within the forests. There are two other groves connected with the history of the redwoods which have escaped scrutiny, mainly because one, the Bohemian Grove, is privately owned and the second, the Grove of Understanding, is not somewhere that ever really existed. Both groves, however, are illustrative of the deep symbolism and potential for ritual and celebration that people see in the redwood forests, much at odds with the much more staid tourism of hikes and camping that takes place in the redwood forests more widely.

It is the Bohemian Grove which offers the most unusual manifestation. As a grove within the forest, it predates the establishment of all the public groves such as Muir Woods as it was privately purchased by the Bohemian Club of San Francisco in 1899. The Bohemian Grove was established as the summer encampment of a San Francisco-based group, the Bohemian Club. After initially only renting the forest, the members went on to purchase a central grove, which was to become the focal point of the site, as well as much of the surrounding land. The artistic and literary membership was quickly superseded by a more wealthy plutocratic dominance, and the activities within the 2,700-acre site gained a mystique and notoriety which will sustain conspiracy theorists for years to come (Figure 6.5).[32]

The Bohemian Grove has hosted American presidents, musicians and actors, is allegedly the site on which the Manhattan Project was conceived, and has been off-bounds to anyone but the exclusively male members of the club for over 100 years. The group gathers for music, plays and

Figure 6.5 Untitled. A campfire in the Bohemian Grove, most likely a photo taken by Richard St. Barbe Baker when he visited California in 1930 on the invitation of Madison. The owl insignia of the grove can be made out on the lanterns hanging from the trees, no date (Baker Papers, University of Saskatchewan).

pagan-inspired ritual such as the Cremation of Care, described by one tres-
passing journalist as a 'macabre, hokey ceremony – with Druidic, Masonic,
Ku Klux Klan, and Aryan forest-worship overtones'.[33] It is a surreal site of
intrigue and excess, yet exposed as having a banality much at odds with its
reputation as the great and the good relax, drink and take in plays, music
and the company of their companions amongst the redwood trees:

> Everything in the encampment is sheltered by redwoods, which admit
> hazy shafts of sunlight, and every camp has a more or less constant
> campfire sending a soft column of smoke into the trees. The walled
> camps are generally about 100 feet wide and stretch back up the hill-
> side, with wooden platforms on which members set up tents. Bohe-
> mians sleep on cots in these tents, or, in the richer camps, in redwood
> cabins. The camps are decorated with wooden or stone sculptures of
> owls, the Grove symbol.[34]

It is likely that the atmosphere was more appealing than the theatrics, with
the quasi-mystical titles of early plays including *The fall of Ug: a masque of
fear*, a dramatic creation that forged a new mythology for its audience. The
redwoods surpassed their role as scenery and became agents in the play, 'red
fingers, nameless in their size and urgency, uppointing him a strange and
certain way of peace'.[35] The Bohemian Club appears in Schrepfer's account
of the redwood's history, as the site of differing approaches to diplomacy
necessary in the League's activities where the Bohemian was seen as too
private and ineffectual a place for such discussions. Their ethos is manifest
in the motto of the society, 'Weaving Spiders Come Not Here',[36] making
the Bohemian Grove a place for recreation and community building, rather
than formal business and politics. The comradery of the Bohemian Grove is
one where the grove has a real role in creating a new collective persona. As
Michael Conan describes, it is a landscape which enables its inhabitants 'to
establish a sense of locality for themselves, thus metaphorically rooting their
own identity in a well-defined part of the material world'.[37]

Richard St. Barbe Baker, the forester and environmentalist, was most
likely invited by one of the founding members of the Save the Redwoods
League, Madison Grant, to the Bohemian Grove in 1930 when the Englishman's tour of America allowed him to travel west and visit the prodigious
trees. The men had things in common beyond their shared interest in trees,
as Baker was a member of the English Eugenics Society and Grant, as well
as being a conservationist, was author of a best-selling textbook on scientific
racism, *The Passing of The Great Race* in 1916. The identity of the leading
Save the Redwood League members as eugenicists is not something that was
widely remarked upon at the time, but something that has become clearer
with time as the interwoven ideologies of wilderness preservationists and
racial hygienists have been more closely examined.[38]

The experience of visiting the grove was one which profoundly affected
Baker and left him sure that he wanted to find one of his very own. It was a

place he wanted to share with his fellow Men of the Trees, to make a place of pilgrimage for tree-lovers around the world and it was on that basis that he went exploring the forests to search for the appropriate place.[39] Writing in his 1943 volume *The Redwoods* he describes his vision for the Grove of Understanding and what potential it held as a gathering place for tree-lovers from around the globe. Baker envisioned it as a public space, though, in contrast to the Bohemian Grove, somewhere where specially commissioned plays could be performed: 'No artificial scenery is needed or desired. Here, among the oldest living trees we might crystallize in a spiritual sense the universal love of nature on the lines of the ancient Greek drama of the past.'[40] It was a place that he described as being a 'Mecca' in an untouched primeval forest, impressing on the reader the scale of the place that he wanted to save.[41]

There are photographs of the Grove of Understanding in Baker's books but no maps or plans are included so it is hard to understand the spatial and experiential qualities of the place: how big it was, how many trees it included and what their distribution was like. But this absence also helps to illustrate two contemporary difficulties in writing and sharing information about both species of tree: firstly, it was problematic to convey the size of the trees in paintings and photographs, with the scale of reproduction allowing nothing more than 'a pygmy representation' of the enormous trees, or necessitating the inclusion of other figurative elements to demonstrate their size[42]; and secondly, there was a great degree of secrecy in which the most attractive groves were earmarked for preservation. Such information could drive up the value of the land whilst negotiations between the lumbermen and conservationists took place and, as a result, Baker was reluctant even to send a map locating the grove in case that imperilled the trees by driving the cost of purchase out of the range achievable by the League.[43]

What remains

Richard St. Barbe Baker returned to England after completing his world tour with a new-found passion to save the Grove of Understanding. An impassioned letter that he wrote from Ceylon, thinking that he was close to dying, sets out some of the new meaning and potency that he had found in the grove:

> I have sojourned in the groves of the oldest living things. I shall bring with me a new message, hope in service and service in hope of man. For true life is service. Trees are forever giving to man. They always give back more than they take from life, and that surely is service.

His contribution to the fight to preserve the redwoods was to take it to an international audience, making the species emblematic in England, at least, of a vulnerable forest giant.[44] The coast redwood forest and the Grove of Understanding were things that he continued to champion in his work, organising frequent visits with members of the Men of the Trees up until the

1980s, long after it had ceased to be known as the Grove of Understanding and became more widely recognised as Stout Grove.

But the grove, whether it is called the Grove of Understanding or Stout Grove, is still important as a totemic symbol in the challenge of bringing about an understanding of the role of redwood forests in the regional ecosystem. It was also a place synonymous with peace, and in 1939 he concentrated his efforts in fundraising for a centre for world understanding which he felt would allow people to connect with all living things.[45] For Baker, finding the Grove of Understanding inspired him to take the fight to save the redwoods to an international audience leading him to be described as the Redwood Saint at his memorial service upon his death in 1982.[46] For Baker, the forest was representative of a much more important environmental process: the life-giving gift of oxygen and the regulation of the earth's atmosphere, and this became his life's work to conceptualise in a way that was emotive and effective.

The story of the Grove of Understanding is one that is lacking in climax. Baker's aspirations for the grove to be a place of pilgrimage for tree-lovers from around the world has not materialised in quite the same way he envisioned. The particular ownership, or designation, of the Grove of Understanding as a place for his society, the Men of the Trees, never took place, nor the plays and music. Yet, that the trees are still there and can be visited is testament to his vision and passion, along with that of all the other people involved in the ongoing fight to save the redwoods. Fifty years after Baker first visited the place he called the Grove of Understanding, he returned to the site when it was included in the 1980 designation of the Jedediah Smith State Park, itself within the Redwood National Park, as an UNESCO World Heritage Site. The Grove of Understanding, as Stout Grove, is today one of the many publically accessible sites in the park near Crescent City and the Oregon border.

The scale of the remaining redwood forest and the groves it contains is testament to the work of many people who fought to keep them in the collective commons. But actions and events outside the forest realm might prove to be as big a threat as logging, as scientists are still seeking to understand the human agency that played and plays a part in the ecology and succession of the trees.[47] This paper has been written at a distance, by an author who has never visited California and only seen these trees in the rather different climate of the United Kingdom. But that is somewhat testament to the emblematic power that they have. At a time of potentially huge environmental shifts, with climate change seeming to be a thing of the present and not of the future, the fate of these huge and quiet sentinels is bound up with our own human future. The words that close Richard St. Barbe Baker's book about the redwoods goes some way to sum up the place the trees, the groves and the forest have in our collective consciousness:

Does it seem strange that an Englishman should concern himself with the salvation of Californian trees? The truth is that I have gained so much inspiration from these groves of immortals that I want them to be handed on for the inspiration of the generations that will follow.[48]

Groves are a powerful metaphor for understanding and communicating the wonder and frailty of the fragmented redwood forests of California. The long-travelled word found some of its most evocative applications amongst the coastal giants, the shafts of sunlight passing between their trunks transporting visitors to other hallowed places and allowing moments of communion with an experience of nature and wonder which is not of the everyday. Used as a means of conveying what was unique about the trees and providing important associations of sacrilege and transgression alongside their destruction, the word grove has attained powerful agency in the preservation and conservation of the redwood forest and the wider forest ecosystem.[49]

Notes

1 Walt Whitman, '253. Song of the redwood-tree', in *A Concordance of Walt Whitman's Leaves of Grass and Selected Prose Writings*, ed. by Edwin Harold Eby (Seattle: University of Washington Press, 1955), Bartleby.com, 1999.
2 James George Fraser, *The Golden Bough: A Study in Magic and Religion* (London: Palgrave Macmillan, 1922, 1974), p. 144–58.
3 Seneca, quoted in John Stewart Collis, *The Triumph of the Tree* (London: Jonathan Cape, 1950), p. 78.
4 Robert Pogue Harrison, *Forests: The Shadow of Civilisation* (Chicago: The University of Chicago Press, 1992), p. 178.
5 Harrison, *Forests: The Shadow of Civilisation*, p. 178.
6 John S. C. Abbott, *The Adventures of the Chevalier De La Salle and His Companions in Their Explorations of the Prairies, Forests, Lakes and Rivers of the New World and Their Interviews With the Savage Tribes Two Hundred Years Ago* (New York: Dodd and Mead, (1875) 1903 ed., pp. 133–34.
7 Paul Kelsch, 'Constructions of American forest: Four landscapes, four readings', in *Environmentalism in Landscape Architecture*, ed. by Michael Conan (Washington: Dumbarton Oaks, 2000), pp. 163–86.
8 Simon Schama, *Landscape and Memory* (London: Fontana Press, 1995), p. 187.
9 Ibid., p. 187.
10 Ibid., p. 188.
11 Kelsch, 'Constructions of American forest: Four landscapes, four readings', pp. 163–86.
12 *John Muir Sierra Club Bulletin*11:1 (1920), pp. 1–4. http://vault.sierraclub.org/john_muir_exhibit/writings/save_the_redwoods_1920.aspx [accessed 28/07/2016].
13 John Muir, *The Mountains of California* (New York: The Century Co. 1894); Sierra Club Vault [website] 2016, Chapter Eight: 'The forests': paragraph three.
14 Muir, *The Mountains of California*, paragraph seven.
15 Schama, *Landscape and Memory*, p. 190.
16 Muir, *The Mountains of California*, p. 189.
17 Elizabeth Stevenson, *Park Maker: A life of Frederick Law Olmsted* (Macmillan: New York, 1977), pp. 247–73.

18 Frederick Law Olmsted, quoted in Reed Noss *The Redwood Forest: History, Ecology, and Conservation of the Coast Redwoods* (Washington: Island Press, 1999), p. 34.

19 Schama, *Landscape and Memory*, p. 191.

20 Susan R. Schrepfer, *The Fight to Save the Redwoods: A History of Environmental Reform 1917–1978* (Wisconsin: The University of Wisconsin Press, 1983), p. 7.

21 Schama, *Landscape and Memory*, p. 185.

22 *Sierra Club Bulletin* 11:1(1920), pp. 1–4. www.yosemite.ca.us/john_muir_writings/save_the_redwoods.html [accessed 28/07/2016].

23 Schrepfer, *The Fight to Save the Redwoods*, p. 7.

24 Ibid., p. 7.

25 These spaces are more part of the parallel story about the establishment of the Redwood State Parks which this paper cannot do justice, but it is described in full in Schrepfer, *The Fight to Save the Redwoods* and in Noss, *The Redwood Forest;* in this paper the focus remains on groves.

26 Schrepfer, *The Fight to Save the Redwoods*, p. 3.

27 Ibid., pp. 3–17.

28 Ibid., p. 13.

29 Ibid., p. 31.

30 Richard Nixon 344 – 'Remarks at the dedication of Lady Bird Johnson Grove in Redwood National Park in California, August 27, 1969' www.presidency.ucsb.edu/ws/?pid=2213 [accessed 20/07/2016].

31 'Save the Redwoods League De'dicate a Redwood Grove' www.savetheredwoods.org/donate/dedicate-a-redwood-grove/ [accessed 27/07/2016] 2016 Save the Redwoods League.

32 Mike Hanson, 'Bohemian Grove: Cult of conspiracy' http://bohemiangroveexposed.com/ (Austin, 2004–2012) [accessed 27/07/2016].

33 Alex Shoumatoff, 'Bohemian tragedy: A guide to the Bohemian Grove', *Vanity Fair*, May 2009 www.vanityfair.com/culture/2009/05/bohemian-grove200905 [accessed 28/07/2016].

34 Philip Weiss, 'Masters of the Universe go to camp: Inside the Bohemian Grove', *Spy Magazine*, November 1989, pp. 59–76.

35 Rufus Steele, *The Fall of Ug: A Masque of Fear* (San Francisco: Taylor, Nash & Taylor, 1913), p. ix, 'Foreword'.

36 Schrepfer, *The Fight to Save the Redwoods*, p. 176.

37 Michael Conan, 'Introduction: The Cultural Agency of Gardens and Landscapes', in *Sacred Gardens and Landscapes: Ritual and Agency*, ed. by Michael Conan (Washington: Dumbarton Oaks, 2007), pp. 3–16.

38 Angus Mclaren, *Reproduction by Design: Sex, Robots, Trees, and Test-Tube Babies in Interwar Britain* (Chicago: University of Chicago Press 2012), p. 1.

39 Richard St. Barbe Baker, *My Life – My Trees* (1970, Moray: Findhorn, 1981), p. 70.

40 Richard St. Barbe Baker, *The Redwoods* (Welwyn: George Ronald, 1943, 1960), p. 134.

41 Richard St. Barbe Baker, *I Planted Trees* (London: Lutterworth, 1944, 1947), pp. 227–28.

42 Schama, *Landscape and Memory*, pp. 195–97.

43 Baker, *I Planted Trees*, p. 222.

44 Mclaren, *Reproduction by Design*, p. 140.

45 Letter from Richard St. Barbe Baker to Mrs Gasque, 20 June 1939. The Baker Papers, the University of Saskatchewan Special Collection and Archives, D. Forestry and Conservation Activities III. Redwoods 1. Correspondence, 1939–82 C/1/2.

46 Edward Goldsmith, 'Address given by Edward Goldsmith at the service of the "Celebration of the life and works of St. Barbe" on 14th July 1982 at St. John's, Smith Square, London', *Trees: Journal of the Men of the Trees* 42: 3 (1982), pp. 12–17 (p. 12).

47 Craig G. Lorimer, Daniel J. Porter, Mary Ann Madej, John D. Stuart, Stephen D. Viers Jr., Steven P. Norman, Kevin L. O'Hara, William J. Libby, 'Presettlement and modern disturbance regimes in coast redwood forests: Implications for the conservation of old-growth stands', *Forest Ecology and Management* 258:7 (2009), pp. 1038–54.

48 Baker, *The Redwoods*, pp. 145–46.

49 The illustrations in this chapter have been drawn from the Baker Papers held at the University of Saskatchewan and are reproduced with their knowledge and approval. The English forester and environmentalist Richard St. Barbe Baker was made an honorary doctor of law by the university in 1971. The author was able to visit the collection there of his personal and professional papers in spring 2016 as a result of grant funding from the Landscape Architecture Foundation Canada and the Canada UK Foundation. It has been assumed that all the photos were taken by St. Barbe Baker, with the exception of the one photo credited to the Moulin Studio, San Francisco.

7 Sacred groves in African contexts (Benin, Cameroon)

Insights from history and anthropology

Dominique Juhé-Beaulaton and Matthieu Salpeteur

In Africa, the expressions 'sacred groves' and 'sacred forests' are commonly used by a wide range of actors, from inhabitants to conservation practitioners and scientists. However, behind these apparently consensual terms, a large variety of situations and phenomena can be found which make any generalization about the characteristics or functions of these sites uneasy and uninsightful. This chapter focuses on case studies drawn from our field experiences in Benin and Cameroon. A close examination of the historical trajectories and of the functions of these sites shows that multiple forms of vegetation cover, of religious practices and ritual prescriptions, and management practices are indeed intertwined in these specific places. In the last decades, some new socio-political dynamics grounded on discourses about tradition revival, biodiversity conservation and cultural heritage have added new layers of meaning and functions to these sites.

History of the notion, use and diversity of the expressions

As far as we know, the expression 'sacred grove' in relation to West Africa was used for the first time by Willem Bosman, a Dutch merchant at the turn the seventeenth century. While visiting the Guinea coast he observed the existence of 'sacred groves' near each village.[1] Many travellers after him also used the term 'sacred grove' or its equivalent in European languages and were interested in these specific ritual places. For instance, in 1933 during the colonial period, Auguste Chevalier, a French botanist, referred to the 'nature sanctuaries of the black people' (*sanctuaires de la nature des noirs* in French). Interestingly, he was also probably the first to express concerns about the future of these sites.[2]

It is only in the second half of the twentieth century that this notion has been used more frequently, in close relation with the raising awareness at an international level of the need to protect nature and to involve local populations in biodiversity conservation. The institutional interest in sacred sites first relates to the Man and Biosphere Program (MAB 1971) and the Convention on Biological Diversity (Rio, 1992). Further programmes specifically aimed at developing conservation initiatives are based on these sites, led by UNESCO: 'Sacred sites – cultural integrity and biological diversity' (1997); 'The importance of Sacred Natural Sites for Biodiversity Conservation';[3] and 'Recognition and conservation of sacred natural sites in protected areas' (2008). Further notions have been integrated into the institutional vocabulary related to heritage and conservation, and underlie new initiatives and policies that deal to some extent with sacred groves. These include the notions of cultural landscape, community forestry or community areas, and more recently indigenous peoples' and community conserved territories and areas (ICCAs).[4] This increasing attention is mainly due to the high value of sacred groves in fostering community-based conservation initiatives; they have long been seen as vernacular conservation sites, where remnants of an ancient forest cover have been protected through religion-based bans.

From the 'nature sanctuaries' of Chevalier to the 'conservatories of fauna and flora' of the botanists of the 1980s,[5] most scientists' perception of sacred forests in Africa have remained much the same: these places are considered as biodiversity reserves which are well protected by religious practitioners.

At the local level, a wide range of names and generic terms are used by the inhabitants of these areas to refer to these sites. In South Benin, for example, quite often the sacred groves are named after divinities: *Sakpata zun* is the name of the place of worship for the Sakpata divinity; and the term '*zun*' means an area with dense vegetation, either bush or forest. *Vodun forest* is the generic expression most used by local people.[6] In Western Cameroon, sacred groves are also named with reference to the tutelary divinity they are dedicated to: *mbing Fo-moughou* is the name of the sanctuary dedicated to the *Fo-moughou* divinity (literally 'the king of sparrowhawks'), *mbing* also meaning bush or forest in the Yémba language spoken in the Menoua district. French speakers usually refer to *bois sacré* or *forêt sacrée* ('sacred grove', 'forest').

This use of the expression 'sacred forest' is problematic, as the two components of this expression, the forest cover and its sacred dimension, are not necessarily components of these sites, and do not always overlap. These sites are more secret than sacred; they often cover large areas within which some places are shrines or temples, and some trees or other living species (plants or animals) are specifically seen as deities. In these cases the expression 'sacred grove' appears to significantly reduce the great diversity of phenomena that it seeks to describe.

Origins, development and functions of sacred groves

In this section we will present examples of wooded sanctuaries in two different contexts: southern Benin and Western Cameroon.

The various populations of southern Benin share a geographical origin (Oyo, a city in the Yoruba region of Nigeria). They gradually moved into this area between the fifteenth and late nineteenth centuries. These groups show many cultural similarities, in particular in language and religion; they

practice vodun cults that include cosmology related entities, ancestors or deified heroes. Some places are feared for various reasons, such as a smallpox
outbreak, violent or suspicious deaths, or supernatural phenomena. Oral
narratives in the region describe the emergence of different kingdoms during
migrations, among them Dahomey, as well as Ketou, Ouidah, Allada and
Porto Novo, which were founded around the seventeenth century. In these
kingdoms, some forests are places involved in the recognition and reproduction of royal powers, and have been home to ceremonies of enthronement
and purification before and after wars, often associated with springs. Others are places of judgement where sacrificial executions used to take place.
They can also be burial places for dignitaries. Some are meeting places for
secret societies that seek to maintain social order and cohesion, while others are reserved to the cult of vodoun, for example *Xevieso* associated with
lightning, *Dan* with rain and fertility or *Sakpata* with earth and smallpox.
Forested sanctuaries are often multifunctional, and it is arbitrary to make
a strict classification between categories because several cults can be hosted
there. They are also places to practise medicine, the initiation of adepts or
even burial. Some are only recognized by a small community at a local level,
while others have a much larger audience among people who have diverse
origins but recognize the same deity (Figure 7.1).

Figure 7.1 Fo-moughou sacred forest, Ntsingla kingdom, Cameroon (M. Salpeteur).

The Grassfields area in Western Cameroon is characterized by a linguistic and cultural homogeneity. Located on a volcanic plateau, it gathers more than a hundred kingdoms which share many common traits including a specific political organization associating typical features of kingdoms (dynastic lineage), a set of councils or secret societies involved in the government of daily affairs, and a symbolic system based on ancestors and tutelary gods of the land. These kingdoms are generally believed to have emerged around the seventeenth century, although their history more likely spans over a longer period.[7] In the Grassfields area, the expression 'sacred groves' pertains to two different systems. In one, the land was not conceived as belonging to humans, but rather as associated with tutelary divinities acting as 'land gods'. A set of shrines dedicated to these land gods (*si la'* in the *Yemba* language) are found at various territorial scales.[8] For example, at the level of the compound, before settling, a group had to find the tree hosting the god of the place with the help of a ritual specialist and make offerings so as to create a bond with this tutelary divinity, who was then recognized as related to the lineage of these first inhabitants.[9] This kind of shrine, dedicated to land gods, is also found at the level of the residential quarter, or of the entire village.[10] Some specific wooded areas, also called 'sacred groves', are closely associated with the political system and found in conjunction with every royal palace. They are a kind of extension of the palace, and host a range of

Figure 7.2 King palace and, behind, the sacred forest. Fondjomekwet kingdom, Cameroon (M. Salpeteur).

ritual and political activities related to the kingdom's affairs. These forests host royal cemeteries holding the skulls of royal ancestors, the houses of governing councils and secret societies, shrines associated with land gods pertaining to the whole kingdom, and tribunal sites, where judgments were made (*cadi*). These palace forests possess a wide range of functions pertaining to various dimensions of daily life and as we will see below, the ways in which these sites are managed in relation to their functions, and their relation to local history, varies considerably (Figure 7.2).

Historical dimensions of sacred groves

In both countries, sacred groves are generally closely linked with a wide range of historical events, from the foundation of villages to specific episodes related to independence wars or more recent political troubles, while population movements and the resulting settlement and resettlement processes have affected the local geography of wooded sanctuaries.

In Benin, when people had to move for reasons such as conflicts following lineage segmentations and other disputes, depletion of farmlands, epidemics, they took their deities with them, resettling them in a new appropriate place. Rituals were performed to check the suitability of the new location, and pacts were made with the local deities and with previously settled inhabitants. In a very similar way, in the Grassfields area the recurring wars between kingdoms and the process of lineage segmentation (that is, when a family group splits to create a set of subgroups) led to the creation of multiple wooded sanctuaries, set up when the royal palaces moved or when a new polity was created through the alliance of groups of different origins. Alliances and pacts often had to be confirmed by the planting of individual specimens of particular species such as *Dracaena arborea* in Benin. The success of the plantation was a good sign for the future of the new settlement. In Benin, the foundation of a new village was always accompanied with the foundation of new sacred places settled near a special tree (iroko, cotton tree, baobab) or inside a grove. This could happen in a pre-existing forest or in a savannah context, inside the village or just beside it. Water springs were also generally associated with sacred places, as the control of water had strategic importance in a country with scarce resources.

The new territory was spatially defined by these religious alliances through their sacred places. In Benin, quite often the name of the locality is the name of the village's vodun or *To vodun*. When a particular event occurred between communities, such as a conflict, the disappearance or death of war heroes or an alliance against enemies, the place where the event took place became a place of commemoration. In many Cameroonian kingdoms, sacred sites host shrines or holy trees that have been set up, or dedicated, to end internal wars relating to succession affairs and local politics, including those initiated after the intervention of the German, English and French colonialists who have ruled the country. Inhabitants still renew their religious pacts, even if, sometimes, the initiating event has been forgotten. A common feature in the two areas concerns the evolving status of kings after death. In Benin, the kings were considered as sacred during their reign and became a divinity after death. Their graves are still places of worship where royal ancestral rituals are performed each year at harvest times. In Cameroon, the royal skulls preserved in 'houses' in the palace forests remain places where offerings to these great ancestors are performed, even after the palace has been moved or destroyed.

All these sacred places associated with trees or groves are places of memory and represent sources for political history. They teach us about relations between populations in terms of migrations, territorial conquests and political domination.

Ritual functions, social dimensions

The wooded sanctuaries are always associated with ritual activities, although in both countries the activity varies according to the type of sanctuary. In sanctuaries dedicated to land gods (*si la'* in Western Cameroon, *To vodun* in Benin), most of the ritual activity is related to health and well-being issues. Local inhabitants and visitors from outside come to these sites to consult the ritual specialists in charge of the sanctuaries, who act as priests of these gods, regarding matters of women's fertility and sickness. Nowadays, problems that cannot be cured by Western medicine, and those related to success in one's life (for example at school or at work) are also dealt with at the shrines. Sanctuaries often host shrines dedicated to several divinities, and each god can be associated with specific issues. Offerings and rituals will be performed by the priests in the name of the visitors. These ritual specialists are differently organized in the two cultural areas. In Benin, adepts called *vodusi* (spouses of the deity, who are females) are led by a *vodunon* ('mother of the divinity', a title given only to a man). In Cameroon, these dignitaries are called *njuissi* (spouse of the god) and *khemsi* (notable of the god), without any hierarchical ranking between them. Besides these common activities, some specific rituals are performed in the most important shrines in connection with the agricultural cycle (for example at the onset of the rainy season or before the harvest). Specific rituals aiming at bringing and re-settling peace are also performed at special occasions (before royal funerals, for instance). In Cameroon's royal forests the ritual activity varies a great deal, in line with the multiple functions of such forests. Rituals and offerings are performed on royal graves and skulls by the king's descent; members of secret societies and councils also perform specific rituals related to their own activities, and some specific ritual cycles are implemented in relation to the initiation processes of the new kings and their fellow nobility. A final important group of rituals which take place in wooded sanctuaries relate to their judicial function. As mentioned above, some sanctuaries specifically host courts and shrines (*cadi* in Cameroon) where local affairs are judged, particularly in relation to accusations of witchcraft. This function was more important in the past, before the transfer of most judicial power to the national states and courts. For instance, ordeals were performed once a year in some kingdoms of Western Cameroon. Even today, some affairs are still judged locally, and some sanctuaries host tribunals following a regular calendar (that is, once a month).

Management: access, prohibitions, prescriptions

Wooded sanctuaries in these two areas are, as we have seen, fulfilling a wide range of functions. The ways they are managed are directly linked with these functions, through a complex set of prohibitions and prescriptions and through practices that pertain to either religious or political dimensions. Some anthropologists have proposed a classification of the wooded sanctuaries based on a gradation of these systems of prohibitions and prescriptions, from simple 'feared place' (for example, a crossroad), where everyday activities are allowed, to places where all human activities are banned.[11]

The management of the sanctuaries deals mainly with two kinds of activity: the collection of resources from the sanctuaries, and the modification of the vegetation cover due to ritual activities and their associated management. In Cameroon, the collection of plant material (fuel wood and medicinal plants, mostly) in the royal forests is, in general, limited to members of the royal family and to the members of councils and societies that have access to the forest. The royal forest is organized in sectors with restricted access, and individuals are allowed to gather resources only in the sectors to which they have access. In the sanctuaries dedicated to the tutelary gods of the territory, the ritual specialists in charge of each site are those who control the takings; they may allow some individuals (for example traditional healers) to collect medicinal plants, or they collect the plant material themselves before passing it on to someone else. Control and restrictions on the gathering of resources can be observed in South Benin too. All the takings in these wooded sanctuaries are performed following a principle of exchange with the tutelary gods: the collector must first bring offerings like salt, palm oil or kola nuts to the gods, and they will, in return, be able to find the plants they are seeking.

In these two kinds of wooded sanctuaries, human activity is associated with specific transformations of the vegetation cover. Clearings are made in front of the shrines so as to make room for the visitors who participate in the offerings and rituals; paths connecting the shrines are drawn through the vegetation, and the people in charge of these sites regularly maintain these spaces (Figure 7.3). In Cameroonian royal forests, houses are built within the forest to host the meetings of the councils. In Beninese *vodun zun,* too, buildings are erected to host gods or their adepts.

Ritual activities also affect the vegetation cover through the planting of specific species that have a ritual function. Tree species such as *Dracaena*

Figure 7.3 Visitor shelter close to the main shrine, Mekoup sacred forest, Bangang kingdom, Cameroon (M. Salpeteur).

arborea are often planted as specimens close to shrines within the groves (Figure 7.4). Botanical and historical studies have shown that these activities, in some cases over a long period, affect the floristic composition and the appearance of these sites.

Archaeology, history and the study of contemporary practices show that many of these sites are not the remnants of ancient forests. On the contrary, they result from either intentional or unintentional actions and management practices that interact with the spontaneous growth of vegetation to give birth to the groves we can observe nowadays.[12] The regular or occasional practices linked with rituals or with shrine maintenance affect the appearance and the composition of the plant cover, for example through clearance in creating firebreaks or open spaces dedicated to rituals and through the planting of trees used as territorial markers like *Newbouldia laevis* or *Dracaena arborea*, a species commonly used throughout Africa for this purpose.[13] The ban on resource extraction that is implemented in many of these sites can favour the survival of relict tree cover as well as the growth of secondary forest on old settlement sites, in climatic conditions allowing the conquest of forest over savannah.[14] The stories of the establishment of these forests show that they should not be assumed to be the remnants of an original forest cover.

Figure 7.4 Dracaena arborea planted in a shrine, Tinou, Department Mono, Benin (D. Juhé-Beaulaton).

Michael Sheridan questioned this point of sacred forests considered as relics, both from an ecological point of view (that is, as remnants of an ancient forest cover) and from a social point of view (for example as sites where unchanged traditional practices are implemented). He named this still dominant paradigm 'the Relics Theory'.[15] For him and for us, the sacred groves are better seen as the result of intertwined dynamic processes with regard to both ecological and social dimensions, rather than as the relics of an idealized past.

Groves in the landscape

In these two countries and in many other geographical contexts too, sacred natural sites are also often associated with salient features of the landscape: a rocky peak, the tops of hills, waterfalls, water springs, caves and so on. They are to some extent embedded and distributed according to the natural features of the landscape. But the political and religious powers exercised by kingdoms and local religious practices have also contributed to the continued creation of sacred places, giving birth to original landscape dynamics over long timescales. When the Dahomeans conquered the coastal kingdoms of Allada and Ouidah in Benin in the eighteenth century, the royal palaces of the defeated kings were abandoned, but a cult of royal ancestors continued and vegetation replaced the palace. It was forbidden to cultivate the holy places and the clay buildings quickly returned to earth. The destruction of the palaces led to the development of a sacred grove which still exists. In Togoudo, the ancient capital of the Allada kingdom, several groves marked the locality of the royal residences, including *Ayizanzun*, the site of a market under the control of a powerful divinity, and *Ahosezun*, the ancient palace of the king's wives who died without children.[16] In Cameroon, similar historical processes can be observed. The history of the kingdoms of this area is full of conquests, annexations and alliances that have directly affected the distribution of wooded sanctuaries. When the palace of a king was threatened, it was moved to a new site, which was chosen according to strategic criteria. On the new site, a new royal forest was planted to host the meeting houses of secret societies, the royal cemetery and so on. An area behind the new palace was left untouched and normal activities banned, so that in a fairly short time the vegetation grew to create a new grove or bushy area. The ancient palace forests, on the other hand, were kept as sanctuaries, as they were home to the shrines of the tutelary gods of the territory and sometimes royal graves. In close step with the history of each kingdom, a process of creation and transformation of wooded sanctuaries took place, which led to the formation of the landscapes we can observe in this area today. Through the study of the wooded sanctuaries it is often possible to reconstruct parts of local history and to identify the successive positions of the kingdoms' capitals (Figure 7.5).

All these sacred sites, which seem to be 'natural' but are actually the outcome of years of human intervention and management, have contributed to shape the character of the landscapes by creating a distinctive vegetation cover and distribution. All the territory was under religious control linked to political power. The plants used for the ritual foundation of cult sites also mark out the fields. They are recognized as land tenure markers because of their religious and symbolic significance. The process of creation and destruction of sacred groves continues through time and gives birth to original landscape dynamics on a long-term timescale.

Figure 7.5 Ayizanzun, the former site of a market in Togoudo, ancient capital of Allada kingdom, Department Atlantique, Benin (D. Juhé-Beaulaton).

Social changes and the dynamics of the heritage industry

These sacred sites are places of power, in both the secular and sacred senses, and because of this social complexity they have undergone many changes since the arrival of Europeans in the region. The spread of Christianity in southern Benin in the second half of the nineteenth century, more recently in Western Cameroon, led to widespread social and cultural change which caused many traditional practices and rituals to be abandoned or transformed, and which weakened the rules controlling access to sacred sites. Some sacred sites are now places of pilgrimage for Christians. The colonial administration was another destabilizing factor. Often the administrators reinforced their positions by naming village chiefs who were not always the holders of legitimately 'traditional' political power. These colonial interventions impacted on local political powers and the sacred sites lost their judicial function with the development of state tribunals. The colonial power ignored the intertwinement of social regulation with religious practices. At the same time, colonial agricultural policies fostered an increase in forest clearance to develop the plantation and cash-crop economy (palm oil, teak, coffee, cotton and so on). The demarcation of forest areas, the building of roads and railways and the forced removal of villages for 'health or political reasons' were all actions by the colonial administration that caused people to scatter and thus many sacred sites were abandoned. In these ways the colonial authorities strengthened the position of the Christian missionaries.

In Cameroon as in Benin, between independence and the 1980s, those aspects of daily life pertaining to 'the tradition' were mostly kept aside by national authorities, in a general effort to aim at modern development.[17] At the same time increased population pressures resulted in greater stress on the environment, in terms of availability of arable land and firewood, to support rapidly growing urban populations.

In Western Cameroon during the armed insurrection that accompanied the independence process (c. 1956–1962), sacred groves were used as refuges by the independence party (UPC) fighters as well as by the local population. This harsh conflict led to the destruction of some royal palaces and associated power symbols, such as palace forests.

In Benin, relations between the state and religious dignitaries were particularly tense from 1975 onwards under the Marxist-Leninist regime of M. Kerekou. The government's fight against the 'backward practices of witchcraft' involved the destruction of many sites by cutting sacred trees and clearing the forests that sheltered the vodun deities. The situation changed during the 1980s, when scientists began to recognize the importance of African sacred forests for biodiversity conservation. Foresters and botanists compiled a national census of sacred forests which was the first step towards the state's recognition of these sites as a part of Benin's natural and cultural heritage.[18] Forest restoration projects were then initiated by the Department of Water and Forests. At the same time, the economic crisis of the 1980s led to the revival of vodun cults. People attributed the crisis to the lack of respect for taboos and the destruction of sacred sites. A number of sites were therefore 'reconstructed' and such activities became common in Benin after the 'democratic transition' of 1989.

From the 1990s onward, local traditional authorities came back into the national political arena, following a process observed in Benin, Cameroon as in other African countries and described as 'the Return of the Kings'.[19] In Benin, the vodun religion received a new status and recognition by the State. In 1993 the Beninese government organized the 'International Festival of vodun arts' at Ouidah, the town from which people were exported during the slave trade. This event led to the recognition of King Kpasse Forest (*Kpasse zun*) as a site of cultural heritage. Since then, this site has become emblematic of the sacred forests of Benin. The religious leader decided to let the tourists visit the forest, a 'living Museum of Vodun'. A monumental gate and a surrounding wall were erected, and guides escorted visitors from all over the world. Other less famous sacred sites in Benin have also been identified as places for ecotourism initiatives, and for the recognition of cultural and natural heritage.

In the Grassfields area of Cameroon, the kings are formally recognized by the national state as administrative auxiliaries and today they still play an important role in the daily life of the local population. However, because they are engaged in a kind of political competition with new economic and political elites, they have set out to strengthen their political power by mobilizing various resources, such as economics (by developing plantations or business), national politics (by becoming members of a national party), or tradition. In this movement the sacred groves, particularly those associated with royal palaces and that embody some of the symbolic features of the traditional role of these kings, receive specific attention from local rulers. In some kingdoms (for example Baleveng or Ntsingla, in the Dschang area), palace forests have been reforested recently with the help of the Ministry of Forests and Fauna, in response to a demand from the traditional authorities themselves to restore 'big and dark forests', as palace forests were supposed to be in olden times, and as traditional rulers expect them to be nowadays. These demands are not linked with biodiversity conservation concerns.

The succession of religious leaders in both countries has become more and more problematic. In the absence of religious leadership, people may lose interest in sacred sites, leading to their destruction and replacement by farmland or public spaces. Increasingly religious leaders have to call on government forest officers to support them in enforcing the rules that protect these sites, so they are gradually giving away control to state agents. This may explain why emerging NGOs specialize in development, environmental protection and conservation work to raise community awareness of the importance of sacred groves for future generations. The actions of the new leaders reflect the influence of the media, education, and rural-urban migration and depend on the interests of the different stakeholders. Nowadays, when they agree to accept their duties and succeed to their ancestor's seat, the traditional leaders are young, literate, often Christian and aware of national and international political trends relating to cultural and environmental issues.

At the same time as these social changes, the two countries have developed environmental legislation following the ratification of international conventions (biodiversity, 1993; climate change, 1994; desertification, 1996). In Benin, there are also plans to protect cultural heritage. The School of African Heritage based in Porto Novo (Benin) used various sacred groves as venues for training and defined management plans for these forests. The concept of cultural landscape, defined in 1992 by UNESCO to accommodate new types of site,[20] can be related to traditional ways of life or religious and cultural phenomena, such as sacred forests. In 2005, the Osun-Oshogbo sacred grove in Nigeria was classified as a World Heritage site, followed by the Kaya forests in Kenya in 2008. In Benin, in spite of the botanists' first census of sacred groves, and the activities of the School of African Heritage, none is yet listed on the national inventory. However, some which possess a particular ecological value are now integrated in the network of Benin's protected areas.[21]

Figure 7.6 Main entrance of Kpasse Forest, a 'living Museum of Vodun', Ouidah, Benin (D. Juhé-Beaulaton).

All these social dynamics have consequences for the exercise of religious practices and the ways in which the sacred groves are managed. As some of these sites are seriously threatened and perhaps destined to disappear, the future of these patches of woody vegetation will depend on how religious practices adapt to prevailing socio-political conditions. In urban contexts, some of these sacred groves are more often considered as places of leisure, particularly in towns like Ouidah where *Kpasse zun* is a living museum and an arboretum, or in Porto Novo, where the ancient forest of the Migan (king's executioner) is now the 'Jardin des Plantes et de la Nature', a botanical garden open to the public (Figure 7.6). Noteworthy sites in other countries include the 'bois de Boulogne' in Ouagadougou, Burkina Faso and the Banco forest in Abidjan (Ivory Coast). Leisure is now added to the religious function of these sites.[22]

Conclusion

At times Africa's sacred groves are still described as natural and cultural relics from a static pre-colonial past, but they are better conceptualized as the result of a dynamic process which is 'simultaneously ecological, social, political, and religious', as Sheridan puts it.[23] They possess a great diversity of forms and functions connected with the history of populations. Between 1980 and 2012, the views of sacred forests shifted from their being places of superstition where witchcraft survived, to places for biodiversity and cultural heritage conservation. Ecotourism has appeared as a new way of valuing and conserving this heritage, both natural and cultural. Nowadays, the growing attention to sacred groves is connected with themes such as biodiversity conservation, heritage recognition or the delivery of ecosystem services (ecotourism, leisure, museum, economic activities and so on). The religious practices which have contributed to the conservation of sacred sites may change under these conditions, responding to changing times as they have always done. If the recognition of traditional religions as part of national heritage could contribute to their transmission to future generations and thus support the conservation of the sacred forests, it could lead to the maintenance of the rituals and practices, part of an intangible heritage which has always evolved through time. The future of the sacred groves seems to depend on social adaptations to the present situation, each according to its particular site.

Notes

1 Willem Bosman, *Nauwkeurige Beschryving van de Guinese Goud- Tand- en Slave-kust*, (Utrecht: Anthony Schouten, 1704), p. 144. William Bosman, *A New and Accurate Description of the Coast of Guinea, Divided Into the Gold, the Slave and the Ivory Coasts* (London: J. Knapton, A. Bell, R. Smith, D. Midwinter, W. Haws, W. Davis, et al, 1705), p. 153.

2 Auguste Chevalier, 'Les bois sacrés des Noirs de l'Afrique tropicale, sanctuaires de la nature', *Compte Rendu de la Société de Biogéographie* n°82 (1933), p. 37.

3 UNESCO, *The Importance of Sacred Natural Sites for Biodiversity Conservation*, Proceedings of the International Workshop held in China, 17–20 February 2003 (Paris: UNESCO, 2003).

4 www.iccaconsortium.org, 'Indigenous peoples' and community conserved territories and areas (ICCAs): A bold frontier for conservation, sustainable livelihoods and the respect of collective rights'.

5 Vincent Joseph Mama, 'Forêts fétiches; modèle de la conservation de la nature en République Populaire du Bénin', in *Surveillance des écosystèmes forestiers et pastoraux*, mars 1985 (Cotonou: Ministère du Développement rural et de l'action coopérative, 1985), pp. 20–4.

6 Dominique Juhé-Beaulaton (ed.), *Forêts sacrées et sanctuaires boisés: Des créations culturelles et biologiques (Burkina Faso, Togo, Bénin)* (Paris: Karthala, 2010), 280 p.

7 Jean-Pierre Warnier, *Cameroon Grassfields Civilization* (Bamenda, Cameroon: Langaa RPCIG, 2012).

8 Matthieu Salpeteur, 'Espaces politiques, espaces rituels: les bois sacrés de l'Ouest-Cameroun', *Presses de Sciences Po* 55:3 (2010), pp. 19–38.

9 Charles-Henry Pradelles de Latour, *Le crâne qui parle* (Second edition of *Ethnopsychanalyse en pays Bamiléké*) (Paris: EPEL, 1997), p. 52.

10 In Benin too, the land used to belong to tutelary gods and the king's rights on the soil were prominent.

11 Stephan Dugast, 'Bois sacrés, lieux exceptés, sites singuliers: un domaine d'exercice de la pensée classificatoire (Bassar, Togo)', in *Forêts sacrées et*

sanctuaires boisés: Des créations culturelles et biologiques (Burkina Faso, Togo, Bénin), ed. by Dominique Juhé-Beaulaton (Paris: Karthala, 2010), pp. 159–83.

12 Peter Geschiere, Witchcraft, Intimacy, and Trust: Africa in Comparison (Chicago: University of Chicago Press, 2013), 322 p.

13 Michael Sheridan, 'Tanzanian ritual perimetrics and African landscapes: The case of Dracaena', The International Journal of African Historical Studies 41:3 (2008), pp. 491–521.

14 Jean-Louis Devineau, Charles Lecordier and Roger Vuattoux, 'Evolution de la diversité spécifique du peuplement ligneux dans une succession préforestière de colonisation d'une savane protégée des feux (Lamto, Côte-d'Ivoire)', Candollea 39:1 (1984), pp. 103–34.

15 Michael Sheridan, 'The dynamics of African sacred groves: Ecological, social, and symbolic processes', in African Sacred Groves, Ecological Dynamics and Social Change, ed. by M. Sheridan and C. Nyamweru (Athens: Ohio University Press; London: James Currey, 2008), pp. 9–40.

16 Dominique Juhé-Beaulaton, 'Sanctuaires boisés: entre histoire et symboles, biodiversité et patrimoines', in Forêts sacrées et sanctuaires boisés: Des créations culturelles et biologiques, ed. by D. Juhé-Beaulaton (Paris: Karthala, 2010), pp. 267–78.

17 Geschiere, Witchcraft, Intimacy and Trust, chapter 2.

18 Nestor Sokpon and Valentin Agbo, 'Sacred groves as tools for indigenous forest management in Benin', Annales des Sciences Agronomiques du Bénin 2 (1999), pp. 161–75.

19 Claude-Hélène Perrot et François.-Xavier Fauvelle-Aymar, Le retour des rois: les autorités traditionnelles et l'Etat en Afrique contemporaine (Paris: Karthala Editions, 2003).

20 Mechtild Rössler, 'Linking Nature and Culture: World Heritage Cultural Landscapes', Cultural Landscapes: The Challenges of Conservation, World Heritage Papers, 7 (2002), pp. 10–15.

21 Projet d'Intégration des Forêts Sacrées dans le système des Aires Protégées (PIF-SAP), (République du Bénin Ministère de l'Environnement. Direction Générale des Forêts et des Ressources Naturelles 2013).

22 Julien Bondaz, 'Parcs urbains et patrimoine naturel en Afrique de l'Ouest. De la période coloniale au cinquantenaire des Indépendances', Géographie et cultures, 79 (2011), pp. 67, 87.

23 Sheridan, 'Tanzanian ritual perimetrics and African landscapes: The case of Dracaena', pp. 491–521.

8 Groves in Chinese gardens and beyond them

Lei Gao

The grove is both a strange and a familiar concept in China. It is strange because it has rarely been discussed in the Chinese context, and there is no straightforward parallel for the word 'grove' in Chinese. The closest one can find is *lin* (林), meaning 'a group of trees'. Constructed from two of the characters meaning 'wood' or 'tree' (木), *lin* means a grove, a wood or a forest, indeed any group of trees or bamboo, no matter what its size and form. One needs to rely on the context or defining words to identify a particular type of *lin*. For example, *shan lin* (山林, in which the *shan* character means mountain or hill) describes a mountain grove, woods or a forest; *sen lin* (森林, in which *sen* means tall and dense with trees, full of trees) means a forest; *sang lin* (桑林, *sang* means mulberry trees) means a mulberry grove, woods or a forest; and *she lin* (社林, in which the character *she* refers to the god of land) means the grove or woods of the god of land.

The grove on the other hand is a familiar concept, especially in gardens. There are several terms meaning 'garden' in ancient Chinese, some of which include *lin*, such as *yuan lin* (园林, garden and grove), *lin quan* (林泉, grove and spring), *lin ting* (林亭, grove and pavilion). A saying which is over 2,000 years old reveals the significance of the grove in garden making: 'a well without a big turtle is a shallow well; a garden without a tall grove is a small garden'.[1]

Looking at English writing about Chinese gardens and landscapes, one finds that 'grove' is a familiar term. In his *Dissertation on Oriental Gardening* published in 1772 (second edition with additions in 1773), for example, William Chambers took five pages to illustrate groves in Chinese gardens and claimed them 'amongst the most interesting parts of the Chinese plantations.[2] Chambers described the appearance of groves, as well as the trees, shrubs, flowers, animals and birds in the grove. He found groves beautiful and useful with flowering fruit trees and game of all sorts, and said 'every farmer may have a garden without expense; and, that if all land-holders were men of taste, the world might be formed into one continued garden, without difficulty'.[3]

Chambers had only been to Canton, in the south-east of China, and his writing is a hybrid of observation and imagination, so we cannot rely on the accuracy of his description, but he provides an interesting departure point for a more thorough discussion of Chinese groves. Apart from the aesthetic and productive meanings that Chambers referred to, are there other reasons why the Chinese might choose to plant a grove in their gardens? How about groves and other forms of *lin* outside gardens – what are they and what do they mean? And might this help us understand why there is not a single Chinese word for 'grove'?

This paper looks at *lin* both in and out of gardens. Various *lin* have been selected from the rich sources in ancient Chinese literature and are grouped into seven sections, based on their type and location. The first four sections discuss *lin* in un-walled landscapes, roughly arranged in chronological order but with some overlapping of time period; the fifth section is about *lin* in relation to Buddhism, both in and outside the walled landscape; the sixth and seventh sections discuss *lin* in gardens. The common ground emerging from this discussion is that where form and scale are the characterising features for the Western understanding of groves, in the Chinese context it is cultural factors which give *lin* their unique character.

The mulberry *lin* (桑林)

One of the kinds of *lin* referred to most frequently in ancient Chinese texts is *sang lin*, the mulberry grove or forest. Mulberry trees have productive functions for silk making and are therefore important in the country that gave birth to sericulture. Silk has been used as an offering for gods, and as a material for making clothes and paper. Mulberry groves have been recommended for planting around homes: Mencius, the disciple of Confucius, said 'Let mulberry trees be planted about the homesteads with their five *mu* (about 1/3 hectare), and persons of fifty years may be clothed with silk.'[4]

The economic importance of mulberry trees has contributed to their cultural significance too, revealed in both mythic and historic stories. In a Chinese legend, the sun lives in the shape of a bird on two huge mulberry trees that have twisted together. In the sixteenth century BC the King Tang of Shang (?-1588 BC, reigned c. 1617–1588 BC) made an important sacrifice in the mulberry *lin*. After conquering the King of Xia, Tang established the kingdom of Shang. However a five-year drought followed, which was interpreted as a signal of a punishment from heaven. Tang went to the mulberry *lin* and prayed, 'I am responsible for all the guilt. Heaven, gods and ghosts, please do not harm my people because of my fault.' Then he cut his hair and fingernails, and was about to use himself as a sacrifice to worship the god of heaven. Suddenly heavy rain poured down.[5] The successful prayer for rain certified Tang's authority, and cleared the accusation that he had killed the previous king.

Apart from praying for rain, mulberry *lin* were also the site of other ritual sacrifices and related social activities. Before the fifth century BC, ritual sacrifices in mulberry *lin* remained a significant social activity which also provided opportunities (sometimes intentionally arranged) for young men and women to meet. So in literature a mulberry *lin* represents a romantic place for lovers.[6] It is said that Confucius' parents had a date in a mulberry grove and later gave birth to Confucius.[7]

One of the important ritual sacrifices held in mulberry *lin* was to *she*, the god of land. Apart from mulberry trees, other types of trees can also form a *she lin*, a grove planted or selected for the god of land.

She lin (社林)

She means the god of land (also translated as god of earth or soil), or a place where the god of land could be worshipped. Because the god of land was responsible for the products growing from the earth, sacrificing to them for a good harvest was an important ritual practised by all circles of Chinese people, from emperors and various levels of governors to individual households and farmers. The *she* used by emperors was called *tai she* (太社, the great altar of the god of land). *Tai she* also has the symbolic meaning of all the land of the country, so having *tai she* meant having the right to rule the country. The *she* used by ordinary people was called *min she* (民社, the folk altar of the god of land). There was only one *tai she* in a country, but many *min she*. It is said in the ancient book *Rites of Zhou* that every twenty-five homes share a *min she*, although every village usually had a *she*.[8] In the second and eighth lunar months every year, people gathered at the *she*, sacrificed to the god of land with goats and pigs, worshipped (in the second month) and thanked them (in the eighth month) for a good harvest (see Figure 8.1). After the rituals were finished, the meat presented as a sacrifice was divided and given to all the participants and households. People then enjoyed themselves with meat and wine, so *she* was also a place for social gathering and celebration. Such traditions lasted many thousands of years before drawing to an end in the mid-twentieth century.

There were different forms of *she*, judging from the documentation. It represents an altar made of earth, with or without a tree on the top. In some *she* the tree was replaced later by a standing stone to represent the god of land. However this was criticised as inauthentic, because it is the living tree rather than the lifeless stone that reveals the earth's power of growing.[9] The tree on the altar differs from site to site. Local trees were the best choice because they had a greater chance of flourishing: pine trees were used in *tai she*, cypress in eastern *she*, catalpa trees in southern *she*, chestnut trees in western *she*, and pagoda trees in northern *she*.[10] In some places, the altar was replaced by a huge tree or a grove called the tree of *she* (社树, *she shu*) or the grove of *she* (社丛, *she cong*), and a temple was often built under the tree to provide a residence for the god of land (see Figure 8.2). Some *she* stand on open land at the edge of the city, the town or the village; some *she* are encircled or accompanied by a group of trees called *she lin* (社林).

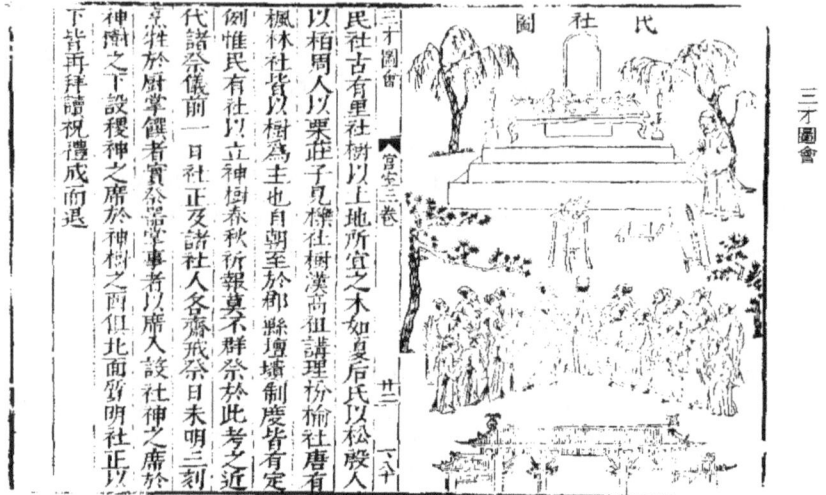

Figure 8.1 A folk altar of the god of land (*min she*) showing an altar of the god of land with different species of trees at four corners. People are lined in front of the altar to worship. The sacrifice is placed on the altar, together with a stele which represents the god of land (*San cai tu hui*, (Collected Illustrations of the Three Realms) a seventeenth-century encyclopedia).

Figure 8.2 The tree and the temple of god of land at the entrance of Guang'ao village, Guangdong province (Ixioino on 28 February 2010, Wikipedia: https://commons.wikimedia.org/wiki/File:广澳伯公1.jpg accessed on 20 February 2016).

Strangely, there is little research on *she lin*, which is almost a forgotten term in China today. Findings from historical texts are limited, so we can only get a rough idea of *she lin*. One of the earliest documentations of *she lin* is from 880 BC: 'In the sixth year of King Yi of Zhou, the King hunted in *she lin*, caught a rhino and returned.'[11] This tells us that this *she lin* was very likely a forest with wild animals, so apart from its sacrificial function, *she lin* was also a hunting park for kings.

A legendary story written down in the tenth century, but said to have happened in the fifth century, has the *she lin* as its setting. It says that a man was told to go to the *she lin*, pick a leaf there and throw it into the river by the *she lin*. A person appeared and guided him to see the river god under the water. The river god gave him a sword, which later protected him from being drown in flood.[12] This illustrates the magical power of the *she lin*, setting it by a river.

An early twentieth-century essay by Lu Xun, 'the *she xi* in my hometown', also described a *she lin*. *She xi* was an opera played to entertain the god of land, usually in the eighth lunar month when the thanks-sacrifice to *she* was held. Lu recalled his childhood experience of going to watch *she xi* in a neighbouring village. They went by boat. When approaching the village, he first saw a grove next to the river. Passing the grove, he suddenly saw the stage of the *she xi*, which was built on the open ground in front of the grove. So we have another grove related to a *she* (very likely a *she lin*) next to a river. The *she lin* is at the edge of the village, so that on entering it, one passes the *she lin* first.

Standing by the river at the entrance to the village, the *she lin* here resembles the *feng shui lin*, which suggests a relationship between *she lin* and *feng shui lin*.

Feng shui lin (风水林)

Feng shui (wind and water) is 'the art-science of siting man-made constructions so that they fit in with the forces of nature rather than clash with them'.[13] The term *feng shui* first appeared in Chinese literature in the fourth century,[14] but the idea of *feng shui* had emerged much earlier. In the pre-Qin (before 221 BC) *Book of Odes* (*Shi jing*) there are descriptions of selecting places according to *yin yang* theory. And in another pre-Qin work, *Book of Change* (*Yi jing*), there is a description of using astrology and geography in order to find the ideal location for a city. *Feng shui* theories flourished in the Tang dynasty (AD 618–907) and became very popular in the times of Ming (AD 1368–1644) and Qing (AD 1644–1911). Mountains and rivers are key natural elements that a *feng shui* master needs to examine when selecting a place for living (such as a city, a village, a house) or the dead (a tomb). If the *feng shui* is not ideal, there are methods for 'correcting' or improving it, one of which is to plant a *feng shui lin*.

Figure 8.3 The water-mouth grove, a type of *feng shui* grove, downstream of the river, where the village entrance is located. Shang Xiao qi village, Jiangxi province (Author, 2003).

There are various types of *feng shui lin*. According to the location, there are village *feng shui lin*, house *feng shui lin*, tomb *feng shui lin*, temple *feng shui lin* and so on, each of which embodies a detailed category. In general, an ideal habitat is sited within the arms of a screen-like mountain range. If there is a gap in this mountain range, or an absence of mountains, a grove or a forest called the dragon seat *lin* (龙座林, *long zuo lin*) is planted to complete the shape. At the place where the river runs out of the village, a grove is often planted by the river to 'lock' good luck in and prevent wealth from leaking out. It is called the water-mouth grove (水口林, *shui kou lin*) (see Figure 8.3).[15]

It is forbidden to damage or cut down trees in *feng shui lin*. This is similar to the attitude towards *she lin*, as both are regarded as sacred places. But there are some differences: the *feng shui lin* helps fulfil a sacred pattern, while the *she lin* is a sacred place in itself. Because the history of *she lin* is older than that of *feng shui lin*, it is possible to use *she lin* as a *feng shui lin*, giving more meanings of *she lin* from a *feng shui* perspective.

Tomb groves

Although creating a benign *feng shui* environment is one reason for planting trees at tomb sites, there are symbolic meanings and 'functional' meanings too.

Planting trees around a tomb had started by the tenth century BC, when an earthen mound was made on a tomb and trees were planted to mark its position.[16] In the western Zhou period (1044–771 BC), the number of trees that could be planted at a tomb depended on the social status of the deceased person, and only imperial families could have trees on their tombs. In the western Han (202 BC–AD 8), lower class merchants and ordinary people were also allowed to have burial mounds and tomb trees, and the rules were more specific too:

> The deceased of high rank has a big grave mound and many trees; while the humble deceased has a small mound and fewer trees; the grave mound of the Son of Heaven is three ren in height (*c.* 5.5 metres), with pine trees planted; the grave mound of a King of State is half the height of that of the Son of Heaven, with cypress trees planted; the grave mound of a senior official is eight chi in height (*c.* 2.2m), with golden rain trees planted; the grave mound of a scholar or lower official is four chi in height (*c.* 1.1m), with ash trees planted. Ordinary people have no grave mound, but willow trees may be planted at the tomb.[17]

By the first century BC, however, differentiation based on social status somehow gave way to differentiation by wealth:

> Ancient people did not make a mound or plant trees at their tombs. They did not sacrifice to the dead in temples; they did not make altars and rooms for the dead to live in, nor did they keep a seat for the spirit of the dead in the temple hall. Later, people started making burial mounds. The common people's mound was half a 'ren' high (c. 0.9m), which is not very high. Today, rich people's burial mounds are like a hill, and arrays of trees are planted to form a grove.[18]

The most common tree species planted at tombs is cypress. It is generally understood that, as an evergreen, the cypress has a meaning of long life, both physical life and the virtue remembered by later generations. However, a legendary story recorded in the third century provides a less well-known reason for this: it is said that this tradition started in the *Qin* dynasty (221–207 BC), when a man accidentally found a goat-like animal from the earth, which was said to eat brains of the dead buried underground. The way to kill such animals was to pierce their heads with cypress branches. Since then, people have planted cypress trees on tombs to protect the dead.[19]

Lin and Buddhism

One often finds *lin* in and outside temples, either planted or naturally regenerated (see Figure 8.4). Apart from the *feng shui* reasons mentioned above, groves have other relationships with religions. Here, I will take Buddhism as an example.

Figure 8.4 The cypress groves in Fayuan Temple in Beijing (Author, 2016).

Buddhism originated in India in the fifth century BC and came to China around the first century AD. There are various schools of Buddhism, some developed in China, but their general principles are the same: to achieve enlightenment and eventually reach a state of nirvana.

A Buddhist temple or monastery is also called a *cong lin* (丛林, grove, thicket), translated from Sanskrit *aranya*, meaning a forest. In the early times of Buddhism in India, monks meditated at the open space in a forest or under trees, from which the name *cong lin* comes.

Planting trees to create a cool and shady place is listed as one of the seven good deeds which give a good return to the people who do them, either in this or in a future life.[20] There are several records showing that when a temple was first built, the master monk planted the same number of trees as the number of words in his selected Buddhist sutra, varying from thousands to tens of thousands of words.

Many temples have bamboo groves. In the Ming novel *Journey to the West*, for example, a purple bamboo grove is mentioned many times; it was where the Bodhisattvatop Guan yin lived and taught Buddhism. In searching for early links between bamboo and Buddhism, I have found two connections: the first Buddhist monastery Kalaṇḍaka Veṇuvana was described as a bamboo garden donated by a converted Buddhist; and in Buddhist teachings, bamboo is an exemplification of Buddha and contains the truth of the universe.[21] Bamboo and trees planted around the temple also create shade and serve to provide storage for timber resources, which can be used in the future for expansion or reconstruction of temple buildings.

Lin in gardens I: a symbol of power and wealth

It is believed that the imperial gardens and parks in China originally contained groves and forests, which also functioned as ritual places and hunting parks.[22] Later, while keeping their functional uses, for example as places for military training and barracks, imperial gardens and parks placed greater emphasis on pleasure and power display. One of the earliest imperial parks of Han times (202 BC–AD 220) was *Shang lin yuan* (上林苑), which literally means 'upper-grove park' or 'upper-forest park', which later became another name for 'imperial gardens and parks'. The making of the park started in 138 BC, on the site of an imperial park of the previous dynasty. *Shang lin yuan* was 300 miles in length, and there were 70 palaces in the park. Each palace could accommodate thousands of soldiers and horses.[23] The trees in *Shang lin yuan* were collected from all parts of China, including rare fruit trees, old trees and over 600 ash trees. A plant list of *Shang lin yuan* recorded over 2000 species,[24] but it is not clear how the trees were arranged: were they groups of single species or mixed species? Were trees planted along the paths or around the palaces? The absence of such information makes it difficult to visualise *lin*, but we can see that what mattered to contemporaries was the symbolic meaning of collecting trees from all over the empire rather than the scale and arrangement of the trees themselves. By having a large collection of trees in his park, the emperor showed that his power extended to all corners of his land.

In Han times, wealthy individuals began to make private gardens. A fifth-century book on eastern Han (AD 25–220) history described a private garden of massive size made by a high-ranked official. In the garden there were man-made mountains with *lin* for the purpose of pleasure:

> The couple Ji and Shou made an extensive park. Earth was dug and piled up to make mountains. Nine hills were created in ten miles of land to mimic the Xiao mountains. There were dark *lin* and deep canyons, which looked like naturally formed landscapes. Rare birds and tamed beasts lived within the garden. Ji and Shou took the chariot car, richly decorated with gold and silver and canopied by a feathered roof, and toured in the garden. They were accompanied by singers and musicians singing all the way.[25]

In this case, it is the collection of rare birds and beasts rather than rare species of trees that reveals the power and wealth of the garden's owners. The *lin* was a place to accommodate birds and animals, creating a forest-like environment in an artificial landscape.

Lin in gardens II: a symbol of nature and self

After the fall of the Han dynasty, there was a period of war and turbulence, and the empire split into many small kingdoms. The size of gardens shrank, but more subtle meanings were incorporated into garden groves. Scholars and literati increasingly valued political and spiritual freedom and promoted a hermit lifestyle under the influence of some great literati such as Tao Yuanming (c. AD 365–427) and the 'Seven Sages of the Bamboo Grove'. Groves in gardens began to symbolize a personal pursuit and to provide a place for a hermit's life. 'There (in the garden) you have groves and rivulets to retreat to, so why would you go to the mountains to seek a hermit's life?'[26]

In his essay, 'Record of the source of the Peach Blossom Grove', Tao Yuanming created a peaceful world behind mountains and peach blossom groves. Villagers there lived a tranquil life without being affected by wars and change of dynasties (see Figure 8.5). Later, the term *tao yuan* (桃源, source of the peach blossom) frequently appeared in literature to show the desire to live in a hidden place with unspoiled beauty and tranquillity. Peach blossom groves are planted in gardens or by water to echo this utopian world created by Tao Yuanming.

The seven sages of the bamboo grove, a description often simplified as 'Seven Sages' or 'the Bamboo Grove', were seven scholars who lived in the late third century. Although not holding the same political views, they all wished to escape from serving the government, which was corrupt and

Figure 8.5 A Ming painting of the source of the Peach Blossom Grove (Peach blossom spring (detail), by Qiu Ying (Chinese, 1494–1552). Museum of Fine Arts, Boston. Accession number 56.494. Photograph © Museum of Fine Arts, Boston).

stifling. They often gathered in a bamboo grove, entertained themselves with wine, art and 'pure conversations' (*qing tan*), a literary style they created which involved witty conversations or debates about metaphysics and philosophy. Their values of spiritual freedom and art rather than material wealth and comfort were appreciated by later scholars, and since then bamboo groves have become a favoured feature in scholar's gardens. As a poet said, 'I would prefer having no meat on the dining table than having no bamboo around my house.'[27] (See Figure 8.6)

Peach blossom groves and bamboo groves are popular features in gardens, while other plants are also used to present personal taste and virtues. For example,

> The landscape of Yong jia is the best under Heaven. There are numerous beautiful flowers and trees growing here. . ., but Zong Qiao only loves citrus trees. He planted a few dozen of them on the islet (in his garden), and enjoyed himself by touring the citrus grove. Because of this, he called himself 'master of the Citrus Islet.[28]

Figure 8.6 A bamboo grove in a scholar's garden of the Ming dynasty (The Stone Table Garden (detail), 1572, by Sun Kehong (Chinese, 1532–1610). Asian Art Museum of San Francisco, Transfer from the Fine Arts Museums of San Francisco, Gift of Mrs. Edward T. Harrison, B69D52. Photograph © Asian Art Museum of San Francisco).

Such passion for citrus trees actually has a historical connection to the great poet and minister Qu Yuan (c. 340–278 BC), which its author pointed out in the latter part of his essay. In a poem called *Ju song* (the Odes of citrus), Qu eulogised citrus trees as faithful and firm – once moved from their homeland, they will not produce fruit.[29] Using citrus as a metaphor, Qu indicated his fidelity to his country, drowning himself when it fell.

Garden essays from the time of the Ming dynasty show that it was common to have single species of trees or bamboo planted to form a grove. Sometimes there was only one grove in a garden, but more frequently several groves were arranged in the garden to create different sub-themed courtyards or spaces. Here are a few of many examples from Ming essays:

> The source of the peach blossom trees is by the left side of the hill. . . . By the right side there is a mound, on which thousands of plum trees are planted. Crossing the rivulet, one comes to the small orchid pavilion, around which there are luxurious groves, tall bamboo, and the winding canal for floating wine cups. The bamboo are as big as a beam, very clean and beautiful. They are so finely polished like the bones of a fan. . . . Peach blossom trees are planted along the river; plum trees are planted in the valley; bamboo are planted to form a grove.[30]
>
> In front of the hall there are clusters of osmanthus trees. . . . Plum trees are extensively planted behind the hall. Beyond the plum trees, there are bamboo groves. Bamboo groves are next to the Buddhist monks' dormitory.[31]
>
> In the middle of the garden there is the main hall. At the right side of the hall, there is a footpath. A few steps from the path, there is a wild plum grove. . . . Strange rocks are arranged to mimic the various peaks in *Pu tuo* and *Tian tai* mountains. On the rock hill several dozens of red plums are planted. Among the plum trees there is a pavilion.[32]

Their cultural meaning is often intertwined with the economic value of groves, such as the mulberry *lin* discussed above. This is also the case for garden groves. For example, the flowers of osmanthus can be used to make perfume and food flavourings; plums can be pickled, dried, candied and honeyed to make snacks, drinks and highly flavoured relishes. These are practical reasons for having groves in gardens. Chambers observed this in the eighteenth century, but these have been largely neglected by modern research on Chinese gardens. It was not until the 1990s that Craig Clunas, in his book *Fruitful Sites: Garden Culture in Ming Dynasty China*, again picked up and highlighted the economic use of garden plants, and claimed that the pursuit of commercial wealth through garden plants played an important role in Ming garden culture.[33] This helps explain why groves of single species of fruit trees were popular at that time.

Paintings also provide images of groves in gardens. In the album of a well-known Ming garden *Zhuo zheng yuan* (拙政园, the humble administrator's garden), thirty-one scenes were illustrated, one on each page. Many of them contain one or more groves, and the majority show groves of a single plant species. Nearly two decades later, the same author selected twelve scenes, redrew them and made another album. Though using the same title, the depicted features are very different in the two versions. For example, in a scene called 'the tent of pagoda trees', the earlier painting showed a man sitting under the shade of three pagoda trees; the later version had eight pagoda trees forming a square enclosure, with three people sitting and talking (see Figure 8.7). When collecting evidence of a grove from pictorial sources, one needs to bear in mind that they are more likely to embody an artistic conception than be a realistic representation.

Figure 8.7 Two versions of the same scene, 'the tent of pagoda trees' in the Humble Administrator's Garden, both drawn by Wen Zhengming (1470–1559). That on the right is an earlier version with three trees of *Sophora japonica*. The number of trees increased to eight in the later version (left). This indicates that what mattered in a designed landscape was not the form, but the sentiments that the form creates (James Cahill, Huang Xiao, Liu Shanshan, *Bu xiu de lin quan* (Garden paintings in old China), (Beijing: SDX Joint Publishing Company, 2012, p. 177). Originals in Metropolitan Museum of Art, New York (left) and Suzhou Museum (right)).

Conclusion

Now we have a better understanding of 'groves' in the Chinese context. *Lin* have economic and aesthetic meanings, but the cultural dimension is the key to understanding them. A *lin* is a place where gods and spirits live (such as *she lin*); a *lin* provides a channel to communicate with superior force (such as the mulberry *lin*); a *lin* fulfils magical patterns to protect good luck (for example, *feng shui lin*). This is also true of the *lin* in Chinese gardens, which were born in forests. Later, when gardens became smaller and further from real forests, a grove was used to indicate the historic connection and the idea of living in nature. Intellectuals and literati also instilled symbolic meanings in groves, which became an effective tool for creating a sense of place. With a grove in the garden, one had a retreat for both body and spirit.

In some cases it is hard to be sure if a *lin* is a grove, a wood or a forest, but such uncertainty does not seem to affect the meaning of *lin*. Symbolic meanings are presented through the cultural significance of a plant, and magical powers are gained in the *lin*. It does not matter how big or small that *lin* is in order for it to have that meaning or power. In historical literature, the size and form of a *lin* is often either not mentioned, or not coherently recorded in the various narratives which survive. We have seen the location of a historically important ritual sacrifice described as a mulberry *lin*, leaving us uncertain whether a mulberry forest or grove was being described, and we have seen two versions of the same garden scene, drawn by the same artist, illustrating very different tree groups. Such discretion or freedom creates difficulty in defining and finding 'groves' in the Chinese context, but on the other hand explains why there is no single word for 'grove' in Chinese. It is the cultural aspect rather than their scale and form that gives importance to a group of trees.

Notes

1 Liu Xiang, 'Tan cong', *Shuo Yuan* (17 BC), vol.16.
2 William Chambers, *A Dissertation on Oriental Gardening*, second edition (London: W. Griffin, 1773), p. 96.
3 Chambers, *Dissertation*, p. 103.
4 Meng Zi and James Legge (trans.), *The Works of Mencius*, Book I, Part I. http://nothingistic.org/library/mencius/mencius01.html [accessed 19/05/2016].
5 Lü Buwei, *Lü shi chun qiu* (247–39 BC).
6 Liu Yuqing and Yang Wenjuan, *Shi jing jiang du* (Interpreting and Commenting the Book of Odes) (Taipei: Long shi-jie, 2014), p. 70.
7 Si Maqian, 'Kong zi shi jia', *Shi ji* (Records of the Grand Historian) (109–91 BC).
8 Xu Shen (c. AD 58-c. 147), 'shi bu', *Shuo wen jie zi* (Explaining Graphs and Analyzing Characters), vol.2. www.shuowen.org/view/64 [accessed 20 April 2016].
9 Qin Huitian (1702–1764), *Wu li tong kao* (a complete research of five rituals), vol. 44.
10 Qin Huitian, *Wu li tong kao*, vol. 43.
11 'Yi wang' (King of Yi), *Zhu shu ji nian* (Bamboo annals) (before 296 BC).
12 Li Fang, 'Shao Jingbo', *Tai ping guang ji* (Extensive Records of the Taiping Era) (AD 977–984).

13 *China: A Cultural and Historical Dictionary*, ed. by Michael Dillon (Curzon Press, 1998), p. 101.

14 Juwen Zhang (trans.), *A Translation of the Ancient Chinese: The Book of Burial (Zang Shu) by Guo Pu (276–324)* (Lewiston: The Edwin Mellon Press, 2004), pp. 58–9.

15 Guan Chuanyou, *Feng shui jing guan – feng shui lin de wen hua jie du* (Feng shui Landscape – The Cultural Interpretation of Feng Shui Forests) (Nanjing: Southeast University Press, 2012), pp. 51–3.

16 Guan Chuanyou, 'Zhong guo gu dai feng shui lin tan xi' (exploring feng shui groves/forests in ancient China), *Nong ye kao gu* (agricultural archaeology), 2002 (vol.3).

17 Zhen Xuan (AD 127–200), Jia Gongyan, Li Xueqin, Zhao Boxiong, *Zhou li zhu shu. Chun-guan zong bo* (Notes and commentaries on the Rites of Zhou) (Taipei: Taiwan Gu-ji Press, 2001), pp. 668–69.

18 Huan Kuan, *Yan tie lun* (discussions on salt-and-iron monopoly), vol. 29 (western Han).

19 Zhang Hua (AD 232–300), *Bo-wu zhi* (Records of Diverse Matters).

20 'Fo shuo zhu de fu tian jing' (Sutra from Budda about all kinds of virtues and benign fields), *Taisho Tripitaka* 16:683. Available online: Chinese Buddhist Electronic Text Association www.cbeta.org/result/normal/T16/0683_001.htm [accessed 20/04/2016].

21 Shi Daoyuan, 'Huihai chan shi' (Huihai Master), *Jing de chuan deng lu* (Inheritance of Buddhist schools and masters) (AD 1004–1007).

22 See 'Ling tai' (Marvellous terrace) in *Shi jing* (Book of Odes), and Ban Gu, 'Yang Xiong zhuan' (Biography of Yang Xiong), *Han shu* (History of Han dynasty) (C1).

23 Author unknown, 'Yuan you' (gardens and parks), *San fu huang tu* (c. Han dynasty).

24 *Xi jing za ji* (essays on Western capital), no.1. http://ctext.org/xijing-zaji/1/zhs [accessed 20/04/2016].

25 Fan Ye (398–445), 'Liang Tong lie zhuan' (biography of Liang Tong), *Hou Han shu* (History of Eastern Han dynasty).

26 Luo Binwang (640–84), 'Chou xi pian', *Quan Tang shi* (complete poems of Tang dynasty), vol. 77.

27 Su Shi (1037–1101), 'Yu qian seng lü zhu xuan' (In the green bamboo pavilion of Qian Monk).

28 Tang Shunzhi (1507–1560), 'Yong jia Yuan jun fang zhou ji' (essay on Mister Yuan of Yongjia's Fragrant Island), *Chong kan Jing chuan xian sheng wen ji* (Reprinted anthology of Mr Jing Chuan), vol. 12. http://ctext.org/wiki.pl?if=gb &chapter=253900&remap=gb#p21 [accessed 2/07/2016].

29 Qu Yuan (c. 340–278 BC), 'Ju song' (Ode of Citrus trees), *Chu ci* (Verses of Chu). http://ctext.org/chu-ci/ju-song [accessed 15/07/2016].

30 Zhang Dai (1597–1679), 'Fan changbai' (garden of Fan Chang-bai), *Tao an meng yi* (recording of dreams in Tao Hut).

31 Wang Xinyi (1572–1645), 'Gui tian yuan ju ji' (essay on back to country life), *Lan xue tang ji* (collected works at Orchid-and-snow Hall), vol. 4.

32 Jiang Yingke (1553–1605), 'Hou le tang ji', *Xue tao ge ji* (collected works at Snow-wave Hall), vol. 7.

33 Craig Clunas, *Fruitful Sites: Garden Vulture in Ming dynasty China* (Durham, NC: Duke University Press, 1996).

9 Korean village groves

Hae-Joon Jung

The notion of the village grove

The *maeulsup*, the Korean village grove, is a wooded area that was planted when a village was established. It defines its space, in front, to the side and behind, and is cooperatively owned, managed and protected by the villagers. It forms a dominant part of the rural landscape.[1] Most village groves are believed to have been planted during the Joseon dynasty (1392–1897), when the Confucian belief system dominated and society largely depended on agriculture. Since then they have mostly been managed by members of family clans who share the same surname. As a result of this remarkable continuity there is often surviving historical evidence about their creation and management, and today they signify one's roots and evoke a sense of nostalgia for the home village. Village groves are seen to reflect the local community, its history, culture and religion. In South Korea, there are about 600 surviving village groves, of various dimensions from 300 square metres to 3 hectares. They are often planted with one species, or a limited number of mostly native trees such as pine trees and zelkova trees which were favoured for cultural and environmental reasons; these species make up an average of two thirds of such plantings.[2]

Maeulsup is a compound word: the prefix *maeul* means a village consisting of a group of houses in the countryside, and *sup*, which derives from the Chinese character *su*, 藪, and refers to 'a marsh or forest with densely grown trees and grass'.[3] So this word indicates that a village grove is closely associated with human settlement. Other words denoting village groves can be divided into three groups according to meaning and function. If a village grove is primarily used to complement the topography or provide shade, the grove is referred to as *magi*, or *jaengi*. When a grove is called *jeong*, or *jeongja*, its main function is that of a rest place for the residents. When a village grove is regarded by villagers as a sacred place housing an indigenous religion deeply rooted in the community which determines the fortune of the village, it is called *dangsup*, *seonghwangnim* or *sillim*.[4] The village grove has long played an essential role, not only sustaining the daily lives of villagers but also promoting a sense of attachment and unity within the community.[5] Modern researchers have tended to explore and intellectualize their ecological benefits, but these would surely not have been recognized as such by the rural communities that conceived them, where self-sufficiency of mental and physical needs was the key objective.

Groves, a threshold between the profane and the sacred

In the traditional belief system, communication with gods is made possible through a medium, sometimes described as the cosmic tree, and in a place which provides the means to distinguish between the sacred and the profane.[6] Repeating the exemplary works of the gods purifies the space. God created a singular universe, but human beings imitated god's creation to create numerous microcosms. As an important component in the materialization of an ideal village form, the village grove formed a threshold between the inner and outer, the sacred and the profane. The village grove provided a restriction and a boundary which separated and distinguished the two worlds and was at the same time a place where the two worlds met.[7] People believed that tutelary gods prevented foreign invasions, evil spirits and the infiltration of diseases through the village grove. When people selected a site to settle, they went through a process of sanctifying the secular space to place it under the protection of god. Only when this was done did people believe that they could be protected from foreign incursions, whether natural or artificial.

Koreans have traditionally considered nature to be a motherly being that takes care of humankind. It follows that they have worked hard, in accordance with the principles of nature, to secure a stable living. Nature has been seen as both that comforting maternal figure and an eternal spiritual essence, so the indigenous religion of Korea has been centred on the worship of nature: life would not be possible unless the principles of nature were respected.[8] In Korea, human life has traditionally been understood in terms of 'horizontal space', as an experience of time passing. However some particular transcendent experiences, such as birth, death or life-change from spiritual enlightenment, are widely related to 'vertical space' as the medium through which the earth is linked to the sky, and humans to Heaven. These experiences of vertical space have centred on groves whose presence is innate in Koreans, a part of their nature.[9]

> In ancient times Hwanin (Heavenly God) had a young son whose name was Hwanung (the son of Heavenly God and the father of Dangun). The boy wished to descend from Heaven to live in the human world. His father surveyed the three highest mountains and chose Taebaeksan Mountain [now Myohyangsan Mountain in North Korea] as a suitable place for his heavenly son to bring happiness to human beings. Therefore he gave Hwanung three heavenly seals and dispatched him to rule over the people. With three thousand of his loyal subjects, Hwanung descended from heaven and appeared under *Sindansu* (the cosmic tree: 神壇樹) on [the top of] Taebaeksan Mountain. He named the place *Sinsi* (the city of god: 神市). He was the Heavenly King Hwanung. He led

his ministers of wind, rain and clouds in teaching the people more than 360 useful arts, including agriculture and medicine, inculcated moral principles and imposed a code of law.[10]

In this case, *sindansu* may indicate not only a single cosmic tree, but also a sacred grove.[11] The myth asserts symbolic identity between the natural features, mountain, tree and grove that the son of the Heavenly God descended to and found his ideal world on earth. This may be the source of the idea of 'the Unity of Man with Heaven', which Koreans see as their contribution to ideas of nature and humanity. For Koreans, villages are the sacred place where human beings coexist with nature. So villagers took advantage of their surrounding nature with respect while territorializing their living places, calling them 'our village'.

At the entrance or inside the grove, people make an altar before a selected tree where the tutelary deity of the village is thought to reside and perform ancestral rites each year. This village ritual was first performed by the settlers of the village for its peaceful sustenance, and in some places it has been passed down to modern times. These rituals played the role of maintaining and strengthening the life of the community. Among them the *dongje*, 洞 祭, was an ancestral rite in rural areas, with the purpose of preventing misfortune and praying for a good harvest or a big catch; it became solidified as a religion with its effect on the lives of villagers.[12] The villagers believed that spirits of their forefathers remained in this place, planted and tended by their ancestors, so this is where the village guardian or ancestral god resides. The villagers perform rituals to this god, praying for a good harvest and to prevent all misfortune. The whole village takes part in the overall process, which includes determining a date during a certain period, picking people to preside over the ritual, performing the ritual itself and eating food together. Nowadays most villages follow Confucian style rituals, but are mixed with shamanistic elements involving spiritual possession through *gut*, shamanic rituals. The ritual has significance not only in itself, but as functions as a self-governing meeting and joint event at which village members can discuss the issues of the village. The village grove embodies indigenous religious elements that began with agriculture, provides a sacred place where rituals were performed, and as a cultural space supported by excellent scenery it was used for rest and recreation, just like modern-day parks.

They were also used as a geomantic ideal where villagers could live in harmony with nature.[13] When the position or topography of the village was not considered harmonious, a village grove was created according to *fengshui* theory to provide energy required for an auspicious life around the village.

Creation of village groves: *fengshui* theory in Korea

Fengshui theory (風水, k. *pungsu*), or traditional geomancy, embodies traditional views of nature and has greatly affected landscape and culture. A set of empirical rules integrate biophysical landscape components with cultural traditions and religious beliefs to guide the practice of selecting and designing homes and burial places.[14] Literally, *feng* means 'wind (風)' and *shui* is 'water (水)'. This suggests that the theory is the traditional philosophy to deal with water and wind. The term *fengshui* first appeared in the Chinese *Book of Burial* (葬書: c. *zangshu*, k. *jangseo*) by Guo Pu (郭璞, 276–324) in the fourth century, where it was explained as follows:

> When *qi* rides with the wind, it disperses; when it reaches water, it ends. The ancients were able to condense the *qi* and keep it from dispersion, to move it and make it cease. Therefore, they called it *feng-shui* (wind and water). The law of wind-water is; getting water is the superior act, hiding from wind is secondary.[15]

Qi (the vital force or energy: 氣, k. *gi*) is invisible, but is believed to pervade every element in nature as the origin of all life. All creatures on the earth, including human beings, are equal and should coexist in harmony as oneness. According to the statement above, *qi* is blown away by wind and accumulated by water. An ideal site attracts little wind and stands near the water.[16] According to *fengshui* theory, *qi* can be distinguished by the shape of land, because invisible and figureless *qi* depends on the earth to flow. The theory shows that the undisturbed flow of *qi* means the maintenance of life.[17] *Fengshui* theory has been a primary means of examining *qi* of the earth, which in turn has the same way of interpreting the natural environment, its form, resources and energy flow. From the examination of *qi* by *fengshui* theory, human beings can determine a use for the site that is suited to its *qi*.[18]

After determining a suitable use for the land, *fengshui* was used to identify the spatial arrangement and structural layout of the selected site. It is believed that if a person lives on a carefully selected site, he or she can benefit from that site and have good fortune. To ascertain whether a site is auspicious or not, four factors are considered: the location and shape of the surrounding hills and mountains; the location, shape and speed of watercourses at the site; the type of person that site is for; and the coordination of that site on a geomancer's compass. The geomantic principles to locate auspicious sites are as follows: 'looking for the dragon (看龍, k. *ganryong*: locating auspicious mountain formation)', 'calming the wind (藏風, k. *jangpung*: finding a place protected from heavy winds)', 'acquiring water (得水, k. *deuksu*: ensuring that water is nearby, but downhill)', 'determining the location of the cave (定穴, k. *jeonghyeol*: a "cave" is not a real hole in the ground but the spot, *hyeol* in Korean, where vital energy flowing through the earth is concentrated and accessible)', 'determining the orientation (坐向, k. *jwahyang*: for this, a geomancer's compass is used)', and 'identifying the shapes (形局: k. *hyeongguk*: determining what objects, especially animals or people, the rock formations surrounding the sites look like)' (Figure 9.1).[19]

Figure 9.1 The guardian tree, or *seonghwangmok*, of Hangchon-ri village in Gimjae Municipal City, North Jeolla Province. This 600-year-old zelkova tree (Natural Monument No. 280) is at the village entrance, by the village shrine and totem poles. The village ritual has taken place every year with prayers for village's prosperity (© Cultural Heritage Administration).

These geomantic practices can be understood as traditional knowledge through which people could acquire suitable residential areas and take measures to enhance the stability of their living conditions. By the examination of the local landscape villagers could believe that they were living on viable land, blessed by the auspiciousness of the site.

Being fundamental to local culture, it is believed that ancient Koreans had their version of *fengshui* to enable settlement within this distinctive environment of mountain ranges.[20] Here the surrounding mountains' function is to tame wind and to gather water, so the setting of most villages is on the southern side of a hill, to catch the winter sun. They are placed well above flood level, with wide fields crossed by a stream and a low hill to the south, sheltering spurs to the east and west, and curving approach roads to block intruders. This traditional style of settlement is called *baesan imsu* (mountain in back, river in front, 背山臨水) and *jangpung deuksu* (protection from wind, obtaining water: 藏風得水) (Figure 9.2).[21]

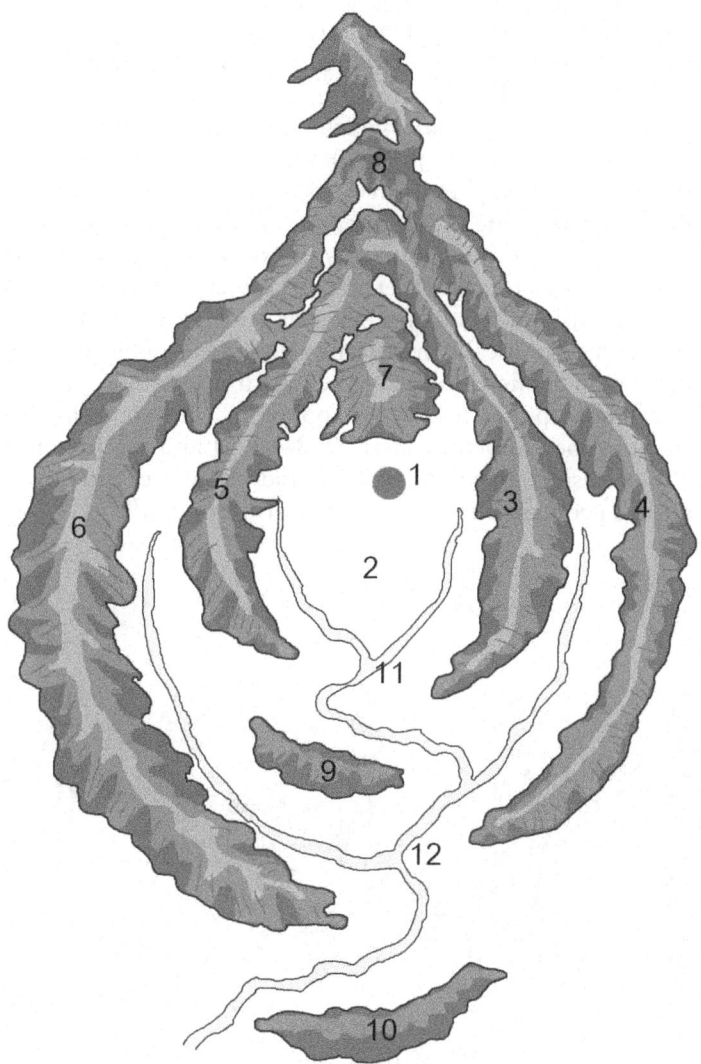

Figure 9.2 The typical topographical pattern of auspicious mountains and watercourses in *fengshui* theory (Author).

The systemization of *fengshui* theory took place under the influence of
Chinese *fengshui*, first introduced to Korea by the geomancer-monk and Zen
master Doseon (道詵, 827–898). In his theories, *bibo* thought (the thought
of complementary, 裨補說) was the central tenet of Korean *fengshui*. *Bibo* is
a method of preparing an auspicious site by making up for elements which

were absent and diminishing strong elements through artificial means. Put differently, if the flow of *qi* through a selected site is too weak or too strong, or the spatial arrangement and structural layout of the site were not sufficient to be auspicious, landscape features around the site would be altered to conform to *fengshui* theory.[22] In Korea this method is called *bibo fenshui* (or complementing *fengshui*), which can be divided into supplementing *fengshui* (裨補風水, k. *bibo pungsu*) and suppressing *fengshui* (厭勝風水, k. *yeomseung pungsu*). The former means that 'good' elements of the sites would be supplemented; the latter means that 'bad' elements would be suppressed. Both methods are means of adjusting the balance of *qi* to prepare an auspicious site.[23] In order to complement geomantic conditions, planting trees to create a woodland belt was widely practised, by building pagodas or temples in critically important places, reinforcing an existing hill or even creating an artificial hill to make up for a weakness. These means of modifying landscape features were applied in cities as well as in rural areas throughout the Goryeo dynasty (918–1392) and the Joseon dynasty (1392–1897). Considerable governmental and private resources were used to carry out this geomantic reinforcement work, or *bibo*.[24]

Figure 9.3 A bird's-eye view of Dalsil Village in Bonghwa County, North Gyeongsang Province. Dalsil is a Korean name, 酉谷 (k. *Yugok*), meaning the valley of the hen. The village was settled in the sixteenth century on the south side of a hill facing the stream (背山臨水: k. *Baesanimsu*), and located in the heart of 'the shape of a golden hen bearing eggs (金鷄抱卵, k. *geumgaeporan*)' in its topography. These are sufficient conditions to be an auspicious site (© Bonghwa County).

Bibo fengshui first appeared in the Buddhist monk Doseon's *bibo satapseol* (裨補寺塔說), which describes a method of supplementing or suppressing geographical energies by erecting Buddhist pagodas on the hotspots of particular sites. The theory may derive from traditional Korean views of nature, which highlighted the balance of the Three Powers, or *samjae* (三才) more: all changes in nature are powered by the combination of Heaven, earth and man. By bolstering the complementary relationship between humans and nature on the site, Koreans believed that a flawlessly balanced unification between Heaven and the earth or *yin* and *yang* could be attained. From this perspective, the purpose of human beings is highlighted as a potent and an essential medium linking both powers. This human-centred view of nature is deemed to be a hallmark of Korean *fengshui* (Figure 9.3).[25]

Sugumagi: village groves for *bibo fengshui* in dwelling places

The desire to live on an auspicious site was widespread across the country, so the use of the methods of *bibo fengshui* to choose and support auspicious sites influenced the selection of residential sites and the management of their surroundings. Complementing and suppressing *fengshui* were then applied in these selected areas in order to tackle deficiencies and to achieve balance and harmony between buildings and their natural surroundings.[26] Yi Jung-Hwan (李重煥, 1690–1756) proposed a theory of environmental space and wrote about the selection of sites for liveable places in *Taekriji* (the book for settlement selection: 擇里志, 1751). In the chapter *Bokgeochongron* (general discussion of liveable places: 卜居總論), geomancy is one of the four important factors that should be considered in the selection of a settlement, along with economic conditions, traits of the villager's mind and natural scenery with beautiful mountains and waters (*shanshui*).

His conception implies that, in deciding where to live, environmental, economic and social soundness should be taken into account in order to make an ideal place and to keep it.[27] As examined in six geomantic requirements for the selection of auspicious settlements, Yi emphasized the importance of the mouth of the watercourse, or *sugu* (水口). *Sugu* is usually identical with a village entrance that is taken to be an important geomantic factor with symbolic meanings, in which water gathers and flows out.[28] With respect to the watercourse, Yi said:

> If *sugu* (the mouth of the watercourse), is warped, organised loosely, empty, or broad, prosperity cannot be extended to the next generation, even if the place has abundant farmland and big houses on it. Those who dwell there will naturally disperse and disappear. Consequently, when people search for and observe a house site, they should look for a stream whose water discharge cannot be observed and a field enclosed by mountains. Although it is easy to find such a watercourse in a mountainous area, it is not easy to find it in a flat land. . . . Whether it is a high

mountain or low land, if water flows nearby in a direction away from the place, it is auspicious. If the place that closes *Sugu* is one layer, it is beneficial; however, if the place consists of three or five layers, it would be much better. This sort of auspicious site can be the place where generations would continue perfectly for a long time.[29]

However, in the real world finding ideal sites matching all these conditions was virtually inconceivable. If the geographical conditions of *sugu* in the selected place were open or wide, it should be blocked and protected with artificial woodlands, ponds or mounds, which was the way to maintain *qi* for the prosperity of the village. These 'layers' are called *sugumagi* (screening the mouth of the watercourse). It was the result of village folk religion and *bibo fengshui* (or the complementing *fengshui*) theory that were combined during the Joseon dynasty.[30] *Sugumagi* is represented by a man-made solid grove belt to control the flow of water and wind; however, it is not a fixed structure like a dam. Instead, by blocking an open space, this artificial landscape feature creates the psychological effect of a barrier.[31]

The *sugumagi* were designed on the basis of one other of the *fengshui* theories for the selection of settlement sites, *jangpung* (protection from wind: 藏風) and *deuksu* (obtaining water, 得水). For this, *bibo* woodland, together with a *bibo* pond depending on site conditions, was created. To close open *sugu*, *bibo* woodland was introduced as a barricade. When this woodland was created in a residential area it functioned like the *maulsup*, the village grove. As *sugu* is usually located in a place where two streams meet and wind from outside is strong, it frequently causes gusts, overflow or soil erosion. Ecological aspects were considered when these groves were created as a countermeasure, aiming for example to prevent flooding during the monsoon season by holding water and firming the soil around the bank while serving as a windbreak forest along a river or coast.[32] In order to obtain and store water and regulate its flow, villagers controlled the shape of the watercourse. If the water flowed directly out of the village, it would be thought that wealth, prosperity, fecundity and abundance could also flow out, so the watercourse would be changed to curve around the village; by creating a woodland belt, such symbolic energy could be blocked from leaving the village and be stored instead. These *bibo* woodlands, *sugumagi*, which screened the mouth of the watercourse, serve not only for adapting to nature, but also through their symbolic function in *fengshui* theory, holding *qi* to maintain the well-being of the village. If a village was located in a sloping area or the topographic shape of a village cannot hold *qi*, an artificial woodland would be planted at the entrance of the village as *sugumagi*. This also aimed to prevent the strong force of fire which might be blown from outside the village, a manifestation of the suppressing form of *fengshui* theory.[33]

In addition to modifying landscape features, symbolic structures were erected at *sugu*, or a village entrance area was created to complement areas that were deficient, whether physically or psychologically. These symbolic features erected in this *sugumagi* area included *jangseung* (a tutelary post), *sotdae* (village guardian poll), altars and shrines. *Sugumagi* with these structures not only played an important role in *fengshui*, but were also sacred places that ruled the village's destiny. In order to protect these features, these guardians of the village, there were *geumsongwanui* (the regulation prohibiting the cutting of pines: 禁松完議) in traditional clan villages.[34]

There is a record of *geumsongwanui*, proclaimed by the Kim clan of *Nae-ap* village in Andong. This village grove was created in a *Sugu* area by Man-Geun Kim (金萬謹, 1446–1500), in the hope of creating a blissful clan village according to *fengshui* theory (Figure 9.4). In order to protect their village grove, the Kim clan's regulation pledged in 1697 that:

> Our ancestors planted pine trees where water flowed out of the village in order to protect the family site and the family graves. Without these pines, it is apparent that there is no *Nae-ap* village. *Nae-ap* village is the place where our clan's ancestral rites have been conducted. The rise and fall of our clan totally depends on these pines. Therefore, you should protect these pines with all your heart, to show that you respect your ancestors.[35]

From the Joseon dynasty, when same-surnamed clan villages commonly appeared in rural areas and considered village groves or *maulsup* to be their most sacred places, the villagers were able to leave them to their descendants.

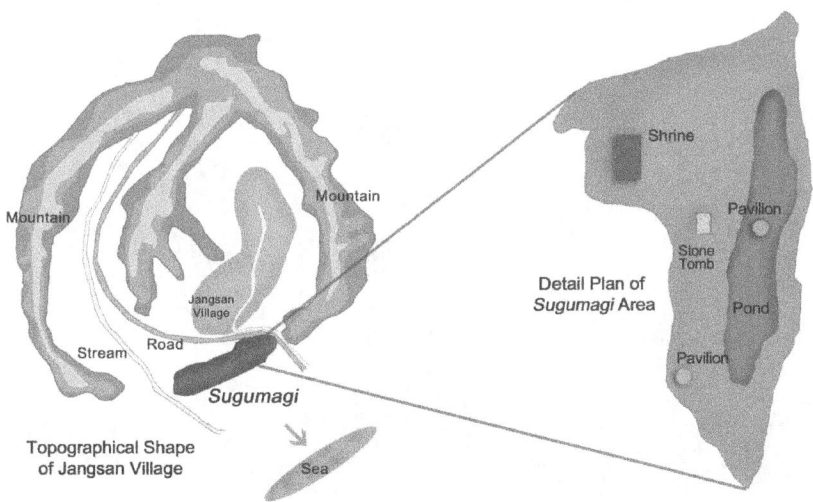

Figure 9.4 Schematic plan of Jangsan Village showing its topography and *sugumagi*, including the grove, pond, a shire, pavilions and a stone tomb.

Village groves have played important cultural and educational roles in promoting a sense of identity, attachment and unity within the community in traditional Korean villages.[36] Although the modes of expression may have changed, it was common practice in fishing, mountain and agricultural villages to conduct communal festivals in these groves, during which time prayers were offered for the continued health and prosperity of their villages. Conducting village rituals such as the *pungeoje* (ritual for big catches: 豊魚祭) and *dangsanje* (ritual for mountain spirits: 堂山祭) in the groves along coastal areas and estuaries played an important role in the forging of traditional lifestyles. These groves were usually situated in *sugumagi* areas, in front of or alongside the village, and created in accordance with *fengshui* theory. In these sacred areas, ritual events of a village, such as *donge* (village rituals) or *gut* (shamanic rituals), and various folk games including *jisin bapgi* (stepping on the spirit of the earth), *ssireum* (wrestling) and *geune* (swinging) took place in order to unite villagers and wish for the village's prosperity. These groves in *sugumagi* areas complemented the energy of the earth or *qi*, and where the village rituals have continued may still be found (Figure 9.5).[37]

As this grove was itself beautiful in appearance, and was created in the most prominent and beautiful spot in the village, pavilions were built there for the appreciation of its scenic beauty, and for communal resting purposes.

Figure 9.5 Jangsan Village Grove in Goseong County, South Gyeongsang Province, is the village *sugumagi*, composed of *bibo* woodland and a pond (© Goseong County).

For the literati of the Joseon dynasty who sought seclusion and retreat, these areas with groves were a place of beauty where the local elite could be united with nature. The aesthetic significance of the *sugumagi* to villages at the time can be seen through the poetry composed by local elite individuals enjoying their natural beauty.

> At the pavilion by the water where an aged pine with graceful foliage droops,
> I, too, am aging like this faithful pine.
> The wish I had to build a pavilion in this beautiful place has finally been fulfilled.
> Time flows with me, and this place is comfortable enough to stay in.
> This beautiful grove has aged with me, and the famous pavilion has found an owner, which is also just right.
> I wish to live here and learn the meaning of the infinite, just as fish and birds enjoy nature.[38]

This poem comes from the name plaque at the *Sehanjeong* pavilion in the Docheon-ri pine grove in Hamyang-gun, Gyeongsangnam-do. It depicts a Joseon scholar who has retired into a rural village and who lives free from worldly cares. In their rural life, local elite people appreciated aged trees and their natural beauty, and took on a sense of transcendence, like that of divine immortals. These *sugumagi* were a concrete representation of this symbolic meaning.[39]

The future of village groves

In East Asian countries, socio-economic and cultural impacts are particularly important for understanding the structure, functioning and dynamics of many of the cultural landscapes.[40] Over the last 50 years, as Korea has experienced extensive urbanization, changes in lifestyle have affected even remote villages. In order to overcome a series of national crises during the twentieth century, including Japanese colonization, the Korean War, military dictatorship and national bankruptcy, there has been an extensive programme of modernization. This was soon considered to be a process of westernization which denigrated local traditions that would tarnish and delay the momentum of economic growth.[41] As a result, Koreans lost their deeply entrenched values, as well as way of thinking and lifestyles and characteristic landscapes; high rise buildings which now accommodate more than half the population were hard to find as late as the 1970s.[42] With the weakening of awareness of values of traditional culture, village groves and their associated rituals have diminished dramatically.

According to studies by Japanese scholars during the Japanese colonial era, over half of all villages performed *dongje*, or village rituals,[43] and after liberation up to the 1960s, the number of *dongje* performed in village groves was considerable. However, the advance of science and the process of modernization meant doing away with superstition, as implemented by Japanese imperialism, while westernization after liberation meant a declining interest in village traditions. As myths and rituals disappeared, the traditional values of the village grove were lost and have faded. This coincided with the structural collapse of rural society. From the perspective of conservation, the village grove faces a gloomy crisis: it has become difficult to maintain them both culturally and physically. Damage to village groves has already reached extremely serious dimensions, and the speed of decline has accelerated.

Fortunately the notion of the village grove has not entirely disappeared from contemporary society. It still opens possibilities for use as an educational resource for history and culture, and thereby serves as a driver to reverse the decline of villages. As awareness of the socio-economic functions of the village grove has increased, central and local governments have lately gradually acknowledged them. Since the 1990s the Cultural Heritage Administration of Korea (CHA) has designated outstanding village groves as Natural Monuments or Scenic Sites, which are types of national-designated heritage, protected by the Cultural Property Protection Act (CPPA). This organization also embarked on 'Heritage resource investigations of village groves', through which those of outstanding academic and landscape value were designated as Scenic Sites, under the categories of 'habitats of fauna and flora with outstanding landscape' and 'famous communities of beautiful plants' (Figure 9.6).[44] These days, most surviving village groves have been designated as either national or locally designated cultural heritage for their landscape value.

The directly related central government departments including the CHA and the Korea Forest Service have devised ways to adopt a well-modulated governance approach, balancing 'carrots, sticks and sermons', offering incentives in the form of subsidies and grants or raising people's awareness to help the managers and owners to make conservation plans that foresee new uses and are sustainable. They embarked on projects like the 'Village Recreation Project' and the 'Direct Payment Programme for Rural Landscape Conservation'. 'The Village Grove Conservation and Management Manual' (2006) and 'the Traditional Village Grove Conservation and Management Guideline' (2008) ensured that local government was given resources for systematically conserving and managing village groves. CHA have recently also provided subsidies to local residents, intending to revive rituals in their village groves. These and festivals run by the local community are expected to encourage tourism as well as benefitting heritage objectives (Figure 9.7). Nongovernmental organizations such as 'Forest for Life' also implement conservation policies to preserve these groves. This organization encourages local residents and outsiders to participate in conservation work in order to strengthen the community spirit. The effect of schemes like this is encouraging and shows that significant results may be achieved through the support of local villagers, rather than large centrally funded initiatives.

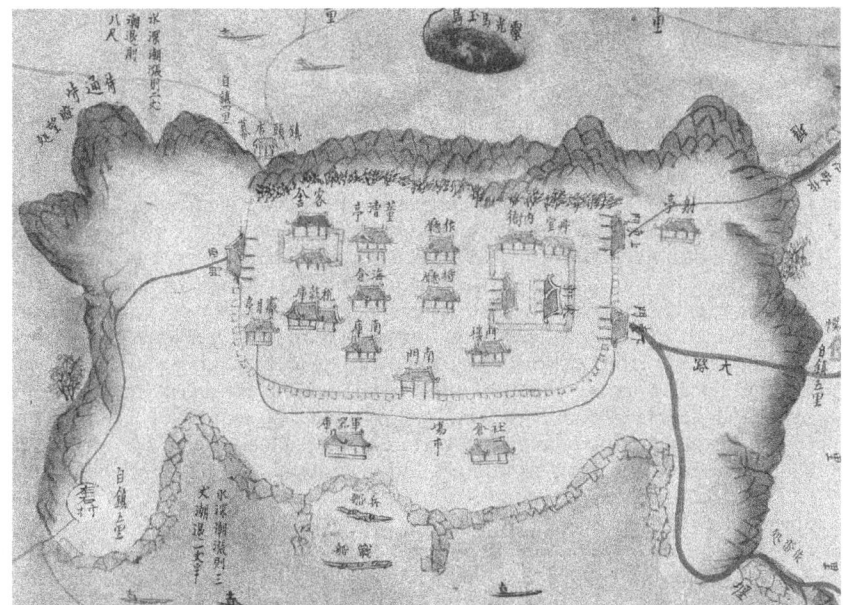

Figure 9.6 One of the groves surveyed was that at the Beopseong garrison now in Yeonggwang County, South Jeolla Province, showing a belt-like grove planted on the mountain ridge behind the naval garrison. This grove is called *Supjaengi*, which means the fortress of grove. Map of Beopseongjin (1872) (© Kyujanggak Institute for Korean Studies).

Figure 9.7 In the grove of Yeonggwang, backpack peddlers who gathered at the fish market commemorated the *Dano*, or the Double Fifth (fifth day of the fifth month on the Lunar calendar), by holding rituals for the dragon king while enjoying boating. These activities have taken place for the past 400 years and developed into the Dano Festival (© Yeonggwang County).

Notes

1 Do-Won Lee, In-Su Koh, and Chan-Ryul Park, *Ecosystem Services of Traditional Village Groves in Korea* (Seoul: Seoul National University Press, 2007). [In Korean with English abstract.]

2 Pine and zelkova trees have been entrusted with symbolic meanings by Koreans. Pine trees, major indigenous species of the Korean Peninsula, symbolically represent honour, fidelity, loyalty, spirit and humbleness. These symbolic meanings reflect the values and spirit pursued by Confucian scholars. Due to their strong growth, dignified appearance and longevity, zelkova trees were called 'Gwimok 樻木', which means 'the tree with a spirit'. With these reasons and beautiful greenery trees made, zelkova trees have been widely used as pavilion trees or village guardian trees, which express wishes for the prosperity and well-being of the village; Hak-Beom Kim, and Dong-Su Jang, *Maulsup, the Korean Village Grove* (Seoul: Youlhwadang, 1994) [In Korean]; Hong, Sun-Kee, and Jae-Eun Kim, 'Traditional forests in villages linking humans and natural landscapes', in *Landscape Ecology in Asian Cultures*, ed. by Sun-Kee Hong et al. (Tokyo: Springer, 2011), pp. 83–98.

3 Hui-Seung Lee, *Great Korean Dictionary* (Seoul: Minjungseorim, 1982) [In Korean.]

4 Hak-Beom Kim and Dong-Su Jang, 'Village grove culture', in Woo-Kyung Sim (ed.), *Korean Traditional Landscape Architecture* (Seoul: Hollym, 2007), pp. 425–457.

5 Do-won Lee, *Ecological Knowledge Embedded in Traditional Korean Landscape* (Seoul: Seoul National University Press, 2004). [In Korean.]

6 Mircea Eliade, *The Sacred and the Profane: The Nature of Religion* (New York: Harcourt, 1959).

7 Won-Suk Choi, *Korean Feng-shui and Bibo* (Seoul: Minsokwon, 2004). [In Korean.]

8 Kyung-Hyun Min, *Korean Gardens*, trans. Halla Kim (Seoul: Borim, 1992), pp. 30–2.

9 Duk-Hyun Kim, 'Traditional views of nature', *Korean Historical Geography* (Seoul: Purungil, 2011), pp. 179–225. [In Korean.] This way of thinking can be traced back to the myth of Dangun, who was the legendary founding father of Korea's first kingdom, Gojoseon (2,333 BC–108 BC), which is recorded in *Samgukyusa* (Memorabilia of the Three Kingdoms: 三國遺事), written by the Buddhist monk Iryeon (1206–1289) in 1281.

10 Ilyon (一然, 1206–1289), The chapter *Gii* (Wonder: 紀異) 1 in vol. 1 of *Samgukyusa* (Memorabilia of the Three Kingdoms, 1281): '庶子桓雄 數意天下 貪求人世 父知子意 下視三危太伯 可以弘益人間 乃授天符印三箇遣佺理之 雄率 徒三千 降於太伯山頂[即太伯 今妙香山] 神壇樹下 謂之神市 是謂 桓雄 天王也 將風伯雨師雲師 而主穀主命主病主刑主善惡 凡主人間三百六十餘事 在世理化'; Ilyon, *Samguk Yusa: Legends and History of the Three Kingdoms of Ancient Korea*, trans. Tae-Hung Ha and Grafton K. Mintz (Seoul: Yonsei University Press, 1971), p. 32; National Institute of Korean History (http://db.history. go.kr) (highlighted by the author).

11 Je-He Im, 'Korean tradition of the mountain worship and the transmission of the mountain spirit worship', in Jong-Seong Kim (ed.), *Mountains and Korean Culture* (Seoul: Sumun Press, 2002), p. 37.

12 Hak-Beom Kim Kin and Dong-Su Jang, 'Village groves', *Journal of Korean Institute of Traditional Landscape Architecture* 23:1 (2005), pp. 145–49. [In Korean with English abstract.]

13 Myeong-Cheol Jeong, 'A study on re-interpretation to function and practice use programme of Rural Community Forest', *Namdo Folk Studies* 15 (2007), pp. 275–313. [In Korean with English abstract.]

14 Ke-Tsung Han, 'Traditional Chinese site selection-Feng Shui: An evolutionary/ecological perspective', *Journal of Cultural Geography* 19:1 (2001), pp. 75–96; Sun-Kee Hong, In-Ju Song, and Jianguo Wu, 'Fengshui theory in urban landscape planning', *Urban Ecosystems,* 10:3 (2007), pp. 221–37.

15 Guo Pu (郭璞, 276–324), the chapter *Neipian* (Inner Chapter: 內篇) in *Zangshu* (the Book of Burial: 葬書, c. 4th–5th century); Juwen Zhang, *A Translation of the Ancient Chinese: The Book of Burial (Zang Shu) by Guo Pu (276–324)* (Lewiston: Edwin Mellen Press, 2004). pp. 58–9; '經曰: 氣乘風則散, 界水則止 ... 古人聚之使不散, 行之使有止, 故謂之風水 ... 風水之法, 得水爲上, 藏風次之'; Chinese Text Project (http://ctext.org/). (highlighted by the author).

16 Ke-Tsung Han, 'Traditional Chinese site selection-Feng Shui: An evolutionary/ecological perspective', *Journal of Cultural Geography* 19:1 (2001), pp. 75–96.

17 Young-Mi Lee and Deuk-Youm Cheon, 'Ecological features appearing in Korean traditional architecture and landscape architecture', *Journal of Architectural History* (2005), pp. 101–15. [In Korean with English abstract.]

18 Du-Gyu Kim, 'Feng Shui (Pungsu): chain of life that connects ancestors with descendants', *Koreana,* 16:4 (2002), pp. 24–31.

19 Kyun Heo, *Gardens of Korea: Harmony With Intellect and Nature*, trans. Donald L. Baker (London: Saffron, 2005); Han-Suk Ock, 'The nature of landscape geomancy and the criteria about selecting the bright yard', *Journal of Cultural and Historical Geography* 17:3 (2005), pp. 22–32. [In Korean with English abstract.]

20 Chang-Jo Choi, 'The characteristic of Korean native *Feng-Shui*', in International Cultural Foundation (ed.), *Korean Feng-Shui Culture* (Seoul: Pagijong Press, 2002), pp. 34–66. [In Korean.]

21 Keith Pratt and Richard Rutt, *Korea: A Historical and Cultural Dictionary* (Richmond: Curzon, 1999), p. 372; Sang-Sup Shin, 'A study on the environmental design principles and space organization of traditional villages', *Journal of Korean Institute of Traditional Landscape Architecture,* 18:1 (2000), pp. 20–31. [In Korean with English abstract.]

22 Woo-Kyung Sim, 'Background of Korean traditional landscape architecture', in *Korean Traditional Landscape Architecture*, ed. by Woo-Kyung Sim (Seoul: Hollym, 2007), pp. 15–56.

23 Du-Gyu Kim, 'Feng Shui (Pungsu): Chain of life that connects ancestors with descendants', *Koreana* 16:4 (2002), pp. 24–31.

24 Hong-Key Yoon, *The Culture of Fengshui in Korea: An Exploration of East Asian Geomancy* (New York: Lexington Books, 2006).

25 Won-Suk Choi, *Korean Feng-Shui and Bibo* (Seoul: Minsokwon, 2004), p. 79. [In Korean with English abstract.]

26 Kim, 'Feng Shui (Pungsu)', pp. 24–31.

27 Sang-Sup Shin, *Korean Traditional Villages and Finding Cultural Landscapes* (Seoul: Daega Press, 2007), pp. 30–1 [In Korean]; Sang-Sup Shin, 'Residential landscape architecture', in Sim, *Korean Traditional Landscape Architecture* p. 95.

28 Chang-Jo Choi, *Feng-Shui: Philosophy of Korea* (Seoul: Mineumsa, 1984).

29 Yi Jung-Hwan (李重煥, 1690–1756), The section on *Jiri* (topography: 地理) in the *Bokgeochongron* chapter (general discussion of liveable places: 卜居總論) in *Taekriji* (the book for the settlement selection: 擇里志, 1751): '凡水口虧疎空濶處雖有良田萬頃廣厦千間類不能傳世自然消散耗敗.　故尋相陽基必求水口關銷內. 開野處着眼. 然山中易得關銷而野中難以固密則必須逆水砂. 無論高山陰坂有力溯流遮攔堂局則. 吉一重固好三重五重尤大吉可爲完固綿遠之基矣.'.

30 Bo-Chul Whang and Myung-Woo Lee, 'Landscape ecology planning principles in Korean Feng-Shui, Bi-Bo woodlands and ponds', *Landscape and Ecological Engineering* 2:2 (2006), pp. 147–62.

31 Hak-Beom Kim and Dong-Su Jang, 'Village groves', *Journal of Korean Institute of Traditional Landscape Architecture* 23:1 (2005), pp. 145–49. [In Korean with English abstract.]

32 Do-Won Lee, In-Su Koh, and Chan-Ryul Park, *Ecosystem Services of Traditional Village Groves in Korea* (Seoul: Seoul National University Press, 2007). [In Korean with English abstract.]

33 Whang and Lee, 'Landscape ecology planning principles in Korean Feng-Shui, Bi-Bo woodlands and ponds', pp. 147–62.

34 Duk-Hyun Kim, 'A study on the rural settlement forest: *Dong-Soo* of traditional village, Nae-Ap in Andong area', *Journal of Geography* 13 (1986), pp. 29–45. [In Korean with English abstract.]

35 Hak-Beom Kim, *Travelling for Korean Scenic Sites* (Seoul: Gimmyoung Publishers, 2013). [In Korean.]

36 Do-Won Lee, *Ecological Knowledge Embedded in Traditional Korean Landscape* (Seoul: Seoul National University Press, 2004). [In Korean with English abstract.]

37 Hak-Beom Kim and Dong-Su Jang, 'Village grove culture', in *Korean Traditional Landscape Architecture*, ed. by Woo-Kyung Sim (Seoul: Hollym, 2007), pp. 425–57.

38 The name plaque at the Sehanjeong pavilion in the Docheon-ri pine grove in Hamyang-gun, Gyeongsangnam-do: '落落松間近水椽, 歲寒心事老林泉, 青山有地遂初服, 白日如年愜晏眠, 嘉樹與人同臭味, 名亭得主訂貪綠, 欲識箇中無限意, 一般漁鳥樂雲川'; Kim, and Jang, *Maulsup, the Korean Village Grove*, p. 124.

39 Kim and Jang, 'Village grove culture'.

40 Joseph Needham, *Science and Civilisation in China* (Cambridge: Cambridge University Press, 1962); Andrew L. March, 'An appreciation of Chinese geomancy', *The Journal of Asian Studies,* 27 (1968), pp. 253–67; Wolfgang Holzner, Marinus Werger and Isao Ikusima, *Man's Impact on Vegetation* (Boston: Springer, 1983).

41 Kyung-Sup Chang, 'Compressed modernity and its discontents: South Korean Society in transition'. *Economy and Society* 28:1 (1999), pp. 30–55.

42 Statistics of Korea, 2006, *Report of Population and Housings* (Seoul: Statistics of Korea, 2006) [In Korean.]

43 Documitsu Nonuyuki, *Groves of the Joseon* (Seoul: Geobook, 2007).

44 Amongst surveyed village groves around South Korea, Seonmongdae Pavilion and Surroundings in Yecheon (Scenic Sites No. 19, designated in 2006) and Beopseongjin Wooded Fort in Yeonggwang (Scenic Sites No. 22, designated in 2007) were designated in this period; Cultural Heritage Administration, *Report of Resources Investigation Research of Maulsup Heritage in Gangwon-Do, Gyeongsanbook-Do and Gyeongsangnam-Do* (Daejeon: Cultural Heritage Administration, 2003).

10 The shrine groves of modern Japan

Yoshifumi Demura

When in 1988 the Japanese film director Hayao Miyazaki produced his popular animated film *My Neighbour Totoro*, a story in postwar rural Japan about the interaction of two sisters with friendly wood spirits, it featured a dense dark grove consisting of huge evergreen camphor trees. Such stands of woody vegetation are generally associated with shrines and can be found not only throughout the countryside, but also in towns and cities, and they often contain a fair range of evergreen broadleaved trees, such as various Japanese oaks,[1] chinquapins and camphor trees. As suggested in the film, the plants in groves were believed to be vegetation surviving from ancient times, a view supported by the main ecological study of plant communities in Japan which highlighted the dominance of broadleaved evergreen trees in the vegetation in the south-west regions of Japan.[2] This is also the most densely populated area, where many groves can be found (Figure 10.1).

These groves are normally referred to as *chinju no mori*; *chinju* means guardian deities of the land while *mori* is translated as grove, and also indicates the shrine itself. Originally, shrine and grove were synonyms, written with similar *kan-ji* characters, 社 (shrine) or 杜 (grove). Despite the fact that shrine groves have existed since ancient times, the earliest usage of the term *chinju no mori* in literature was in Tayama Katai's 1902 novel *Jūemon no saigo* (The end of *Jūemon*) since when it has come into common usage.[3] This novel promoted the assumption that the age-old groves had remained constant, but recent ecological research suggests that they were shaped primarily through natural succession after the Meiji period (1868–1912).[4] Stimulated by a relatively warm and humid climate the present appearance results from an unprecedented period of almost half a century in which these groves have been virtually left untouched. The absence of traditional management of vegetation was based not only on a Rousseauesque naturalism, but was also the result of neglect after World War II, when groves had been venerated as an embodiment of the Imperial institution. With their provenance disguised by recent prejudice, few people are aware of the real

Figure 10.1 An example of a grove at the confluence of two rivers within an urban area: *Tadasu-no-mori* at the Shimogamo Shrine, Kyoto, Japan (Yoshifumi Demura).

history of such groves as part of the modern political process aimed at establishing a democracy in Japan.

This chapter concentrates on the modern era and investigates the significance of groves as symbols of the modernization of Japan, considering how these groves have come to be regarded as ancient vegetation.

Social and political context

The shrine in Japan is a Shinto institution dating from the early eighth century that incorporates the worship of ancestors and natural spirits and a belief in the sacred powers (*kami*) present in both animate and inanimate things.[5] Shrines have traditionally provided an important focus for the communities responsible for them, so many local traditions attach to them. However in the dramatic modernization initiated by the Meiji Restoration of 1867, an oligarchic leadership was established that gave power to a handful of men who aimed to restore central Imperial control with Shinto as the state religion. This attempt to revive Shinto as a national faith failed as a result of interference from Christian global powers who spoke in the name of 'religious freedom'. Instead Shinto became a 'non-religious' patriotic institution supporting daily life and was re-presented as a national tradition; the government established the Shrine Bureau in the Home Ministry in 1900.

While Shinto shrines vary according to local traditions, they share common elements (Figure 10.2). A *Torii* gate provides a threshold to the sacred precinct which contains the main building and the house for worship. This is symbolically guarded by *tamagaki*, sacred fences with a plain structure. Inside these fences, the main path leads to the inner part of the precinct, but it is usually not arranged on a linear axis. The house of worship, in which people perform rites to pray, is usually situated in front of the main shrine building, though there are examples of shrines without a main building, where a grove or a mountain is regarded as the focus of worship. The shrine precinct is necessarily surrounded by the grove, though this does not usually have any direct function related to the rituals of the shrine (Figure 10.3). Although some received national support until the end of World War II, they have mostly been looked after by organizations of villagers or local communities. Since the end of the war, the groves have been maintained by these local organizations, some with support from modern landscape engineers, some by local people and some just left as they were.

Figure 10.2 Typical shrine elements and their arrangement (drawing: Yoshifumi Demura).

Figure 10.3 The *Torii* gate entrance to the Hachiman Shrine in Ena (Yoshifumi Demura).

The shrine merger measure in the modern period (1906–12)

It is said that the early Meiji government led by oligarchic leaders mainly embraced three aspirations: to accomplish industrial revolution, to develop national defence and to establish a limited democratic system which would lead the nation from the top.[6] Although these aspirations were not realized at the time, they remained the aim of the government in the later Meiji period after 1900. Yamagata Aritomo, a survivor from the early leadership of the oligarchy, institutionalized the system of local government in 1888 as the first Home Minister, presenting a vision that important local figures like wealthy merchants and landlords should help local communities to sustain the national body, in a form of top-down democracy. He thought the local leaders could run local government by themselves because of their wealth and they would be willing collaborators with the national administration.

Local development, including the construction of infrastructure for industry, agriculture and forestry, traffic, prevention of disasters and education, was politically focused at the beginning of the twentieth century, reflecting the impact of the new National Diet and party politics supported by the landlord class. This local development can be viewed from a different standpoint, as a countermeasure to party politics, which tended to focus on finance favourable to the parties' own supporters. The idea was to set up a form of local government in which ordinary people would help each other under the guidance of local establishment figures. This notion was embraced by the group known as the 'Yamagata faction' in the Home Ministry, which fostered a new generation of bureaucrats who tried to lead the people with an inexhaustible sense of mission. This group included Tohsuke Hirata and

Rentaro Mizuno, who were succeeded by Tomoichi Inoue and then Kazumi Iinuma.[7]

The Provincial Bureau in the Home Ministry wanted to reform the system of local autonomy to hold the nation firmly together, particularly during the economic confusion immediately after the Russo-Japanese War (1904–05), which had exhausted Japanese resources. Financial problems were severe in Japanese society at this time, and an Imperial Rescript called '*Boshin* Rescript' was promulgated in 1908. It ordered that people should strive for frugality and work hard to strengthen the national body. The Home Ministry promoted the Local Improvement Movement in line with the Rescript. The elite bureaucrats had a strong sense of responsibility and wished to construct a local autonomous structure to manage modern society. Inoue explained this by saying, 'Japanese national and local governments should maintain good relationships or, rather, the national government is willing to care for local governments as if it were tending to a baby'.[8]

These bureaucrats shared the idea that shrines should become central places for rural communities. The shrine would act as a part of the village structure, a symbol in traditional form of the villagers' ethics and reverence, and be placed right in the centre.[9] They believed that the village would stand as a unit with its own autonomy, rooted in the traditions which hold communities gathered around shrines, set apart from the central political struggle.[10] In 1906, the Japanese government initiated a widespread merger of shrines, advocating 'one shrine per village', by implementing two measures.[11] First, shrines were selected for the granting of ceremonial offerings; they received a subsidy provided by the government (Imperial Ordinance No. 96, April). The ordinance required the shrine to have accounts which would demonstrate that the Shinto rites were being managed financially. Another ordinance allowed the shrines to acquire properties on their merged sites to create capital (Imperial Ordinance No. 220, August). As a consequence of these ordinances, the number of shrines decreased from 193,298 in 1903 to 122,593 in 1914.[12] That is, over 70,000 shrines were demolished.

The results differed from prefecture to prefecture, because implementation of the measures was left to the governors. Terrible deforestation occurred in Wakayama Prefecture because of this movement. Kumagusu Minakata (a folklorist and authority on myxomycete in Wakayama Prefecture) resolutely opposed the government's measure in 1909 after observing these disgraceful operations, and in the end he stopped the shrine merger policy. He criticized the mergers, focusing on the collapse of the shrine groves in various ways (see below). He used the newspapers, asked a member of the Diet to pose questions in the National Parliament and sent a personal message for the Home Minister. According to his observations, one of the driving forces behind the mergers was the enthusiasm of local governments to increase their capital by cutting down and selling the trees from the merged shrine groves, in line with Imperial Ordinance No. 220. He pointed out that the extra money from the sale of the wood did not compensate for ruining the villagers' shared places of enjoyment, including those provided by the groves. He compared the traditional groves to chapels or churches built in parks in European countries, there to heal people's spirits and to create a sense of sanctity.[13]

In 1912, Minakata contributed to the magazine *Japan and Japanese* to explain why he opposed the measure. He said that the shrines were in a critical situation and enumerated eight risks of losing them.[14] He argued that the notion that a shrine merger would bring the people to a reverential state of mind was the height of folly; the government was being deceived by reports written by local public servants. His first point was to combat the government's measure by expressing his approval of the government's objective. He said that at the bottom of the political system 'wicked persons' were causing deforestation through their selfish desires. The government left the decisions about whether to implement the measure to the prefectural governors, and they left it with the chiefs of county villages 'who were only interested in writing reports of their own achievements'. Regardless of the essential importance of the various shrines, they ordered them to merge their assets.

Minakata then suggested that the shrine mergers tended to remove the groves and therefore the existence of the deity connected to the land, which was important for local communities. He said that the mergers prevented harmonious relationships and therefore prevented self-governing bodies from working effectively. Merging shrines decreased local vitality, he argued; it deprived people of healing resources, weakened their solidarity and hampered ethical development. The mergers lessened patriotism and damaged public peace in an area, reducing its benefits.

Minakata explained the virtue of the Japanese shrine style in which 'dense divine groves with evergreen trees' were more important than the shrine's buildings. He considered that the government was influenced by an admiration for Western culture, in aiming to establish a smaller number of gorgeous shrines. Instead of magnificent and durable architecture like that in Western churches, which were made of solid materials, 'Japanese shrines had groves, comprised of exceptional large old trees and rare species'. He insisted that the secret reason why the old and simple shrine buildings were regarded as sacred was their proximity to the surrounding 'primeval' forests. He was conscious of the value of the dense, ancient groves surrounding the shrine buildings and stressed it.

The last two points stemmed from Minakata's own professional experience, when he observed that the merger of shrines destroyed historical sites and traditions, and destroyed the natural landscape and its precious natural treasures. He said that the precious rare plants had survived because shrine groves were the place where people had offered them in tribute to the gods. This notion was part of the emergence of a movement for the preservation of natural treasures. The sense of loss of these precious species and historical sites joined with a movement to preserve historic sites and sites of scenic beauty and eventually led to a law for the preservation of cultural property, the Historical Spot, Scenic Beauty and Natural Monument Preservation Law of 1919 (hereafter, the Preservation Law).

We should note that he characterized the value of the shrine groves as places which had preserved local native trees 'for hundreds and thousands of years'. He presented several practical examples to explain how such precious groves disappeared in shrine mergers. He argued that the shrines were worth preserving and researching in order to understand their 'natural vegetation' and their relationship with Japanese land history, quoting what Seiroku Honda, an eminent forester, had stated the previous year. From recent study it is known that the vegetation of shrine groves in his time was a consequence of the forestry policy of the early Meiji government, a recoil from excessive deforestation, which ordered the nation not to cut any trees in mountainous sites to protect the soil.[15] Many shrine groves have not been sustained for thousands of years, but rather had grown into dense uncontrolled forests in those 30 years. The appearance and loss of the overgrown groves must have evoked a sense of preciousness for contemporary intellectuals. This was the period in which the modern image of groves was established.

Creation of the Meiji Shrine grove

The value of the shrine grove which emerged in the struggle around the shrine merger policy can be seen in the story of the construction of the Meiji Shrine's huge, 72 hectares grove (1915–21) (Figures 10.4 and 10.5). One official said that 'with swelling national sentiment, the construction of the Meiji Shrine seemed a turning point to enhance the zeal for shrine reverence among the Japanese people'.[16] It was also a technical trial to create an ideal grove in a metropolitan area from scratch. This project was the incubator which finally added 'eternal dense divine shrine grove' and new imagery of broadleaved evergreen trees to the popular view of groves.

Immediately after the Meiji emperor died on 30 July 1912, it was decided that the site of the mausoleum containing the emperor's grave should be in Kyoto rather than Tokyo.[17] With a great sense of loss, influential figures around the Tokyo Chamber of Commerce started an invitation movement to construct a shrine named after the Meiji emperor in Tokyo, in place of the mausoleum. They thought that there should be a memorial to the emperor in Tokyo to sustain its citizens' honour, as Tokyo had been the capital of the empire since the Meiji Restoration, and if they could not have the tomb and mausoleum, they could create a shrine instead. The reign of the Meiji emperor corresponded with an epoch of upheaval that they had been through together. They yearned for a physical place to pray and share veneration. The concept of the shrine embodied two elements from the outset: an Internal Precinct (内苑) and an External Precinct (外苑) to illustrate sacredness and commemoration respectively (Figure 10.6).[18] The Internal

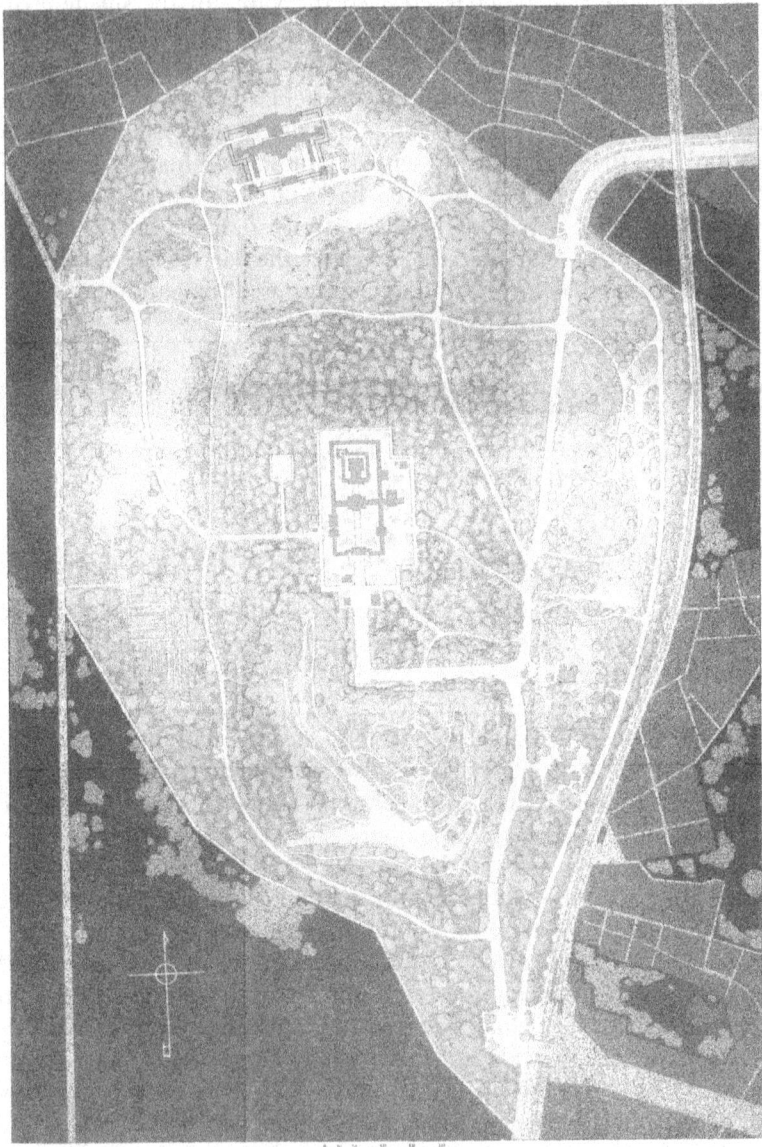

明治神宮境内平面圖

Figure 10.4 Plan of the Internal Precinct of the Meiji Shrine (Appendix to The Bureau of the Meiji Shrine 'Commemorative Issue of the Meiji Shrine' (1923)).

Figure 10.5 The grove within the Internal Precinct of the Meiji Shrine (Yoshifumi Demura).

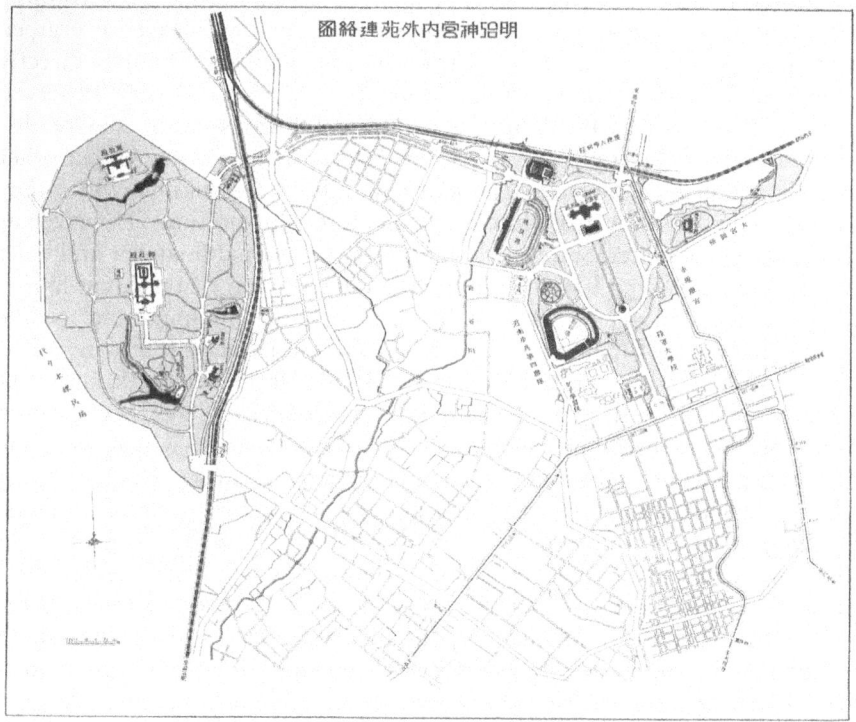

明治神宮内外苑連絡圖

Figure 10.6 Dividing the functions of the Meiji Shrine created two interrelated parts, the Internal Precinct (west) with the shrine grove, and the External Precinct (east) with memorial facilities, connected by a tree lined road (Service Association of the Meiji Shrine, 'Commemorative Issue of the Meiji Shrine External Precinct' (1937)).

Precinct was imagined as a huge and dense divine grove, using terms like shin-gen (森厳), meaning a dense solemn forest.

The decision to go ahead with the construction of the Meiji Shrine was made at the Cabinet meeting on 28 October 1913, and on 20 December the Research Council for the Shrine Dedication (hereafter, the Research Council) was established under the Home Minister, who acted as chairman; its members were political or financial notables, home bureaucrats such as Mizuno and Inoue, and experts in forestry, architecture and literature. This Research Council continued preparing the master plan until the Meiji Shrine Construction Bureau was established on 30 April 1915.[19] Seiroku Honda, a prominent doctor of forestry and a member of the Research Council, began to plan the shrine site in Yoyogi (the present situation) with his students, Takatoku Hongo who was a lecturer and Keiji Uehara, even before the site decision had been made. The three of them designed the entire new huge grove, sharing a concept for the shrine site and that it should be 'the eternal forest'.[20]

At first, Honda and his group thought the 'eternal' forest must consist of coniferous trees because they were evergreen and would make a broad shade which would evoke a deep emotion of reverence. However, the group recognized that the Tokyo site was unsuitable for coniferous trees from the view of forest ecology, but that it was suitable for broadleaved evergreen trees. This recognition came from the knowledge they had obtained directly or indirectly from modern German forestry.[21] It had also been observed that conifers were easily damaged by the sulphurous acidic smoke discharged around metropolitan areas like Tokyo. They recommended other sites inland, such as the base of Mount Fuji, Nikko or Tsukuba, which they thought would suit the growth of dense coniferous plantations. However, as Honda confessed later, Shibusawa, who was the leading financier and chief of the volunteers who would support the construction of the Meiji Shrine in Tokyo, pleaded personally with Honda to change his mind and to take up the challenge of making a great landscape of artificial forest in Tokyo which would equal a natural forest. So Honda decided to seek a way to create a huge grove in plain wasteland.[22] The concept which they adopted then was the natural regeneration of broadleaved trees, which were the natural vegetation of this plain in the middle part of Japan. That is, shade-tolerant broadleaved trees would be planted, not narrow-leaved conifers. They would grow fruit and drop their seeds. The trees shut out the light and wind, and the dead leaves become compost, which helps keep the soil fertile. They would grow up and be successors to the mother trees, so the forest would last forever, Honda and his group thought, so the shrine would be enveloped in an evergreen broadleaved forest, mainly consisting of species tolerant of smoke, such as Japanese oak trees (樫), chinquapins (椎) and camphor trees (楠).[23]

Hongo explained their choice by saying that the existing trees should be used as elements of scenic beauty, that the donated and newly planted trees would be arranged into a suitable grove landscape for the shrine and that

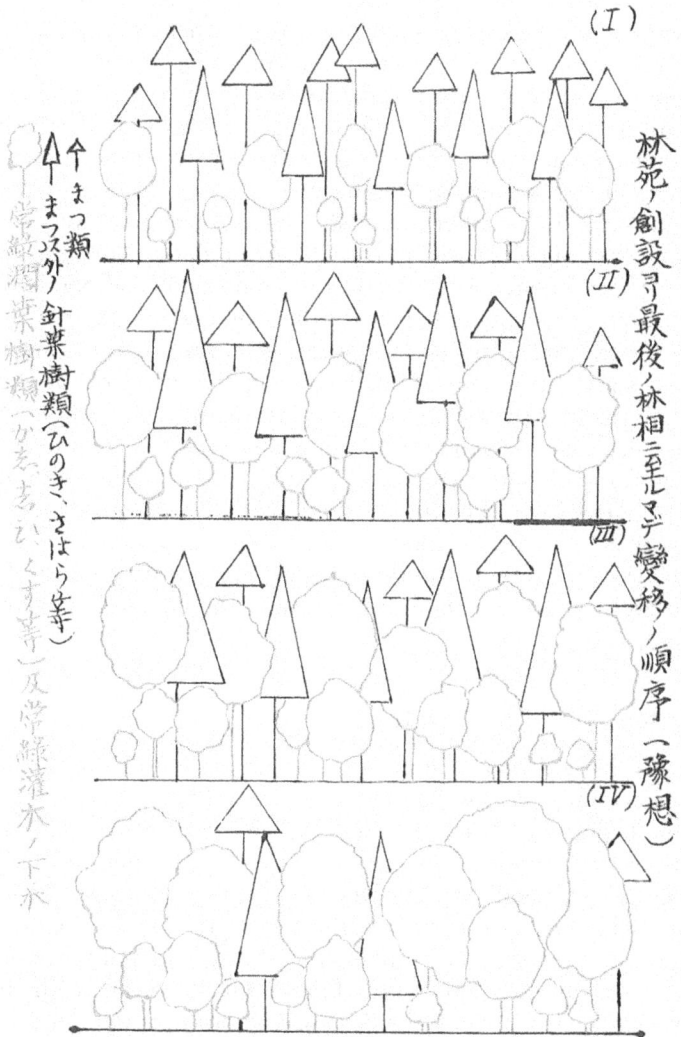

Figure 10.7 Schematic diagram illustrating ecological succession: 'The order from the creation to climax vegetation of the shrine grove (prospect)'. Key: △ (triangles) = Pines; △ (long triangles) = Conifers other than pines; ○ = broadleaved evergreen trees (Takatoku Hongo, *The Plan for the Meiji Shrine Grove* (Tokyo, 1921), p. 53).

the grove should be designed to transform gradually into an ideal forest physiognomy. He explained this process, an idea influenced by the ideal of German forestry, especially Karl Gayer and his follower Heinrich Mayr. He used a sketch to show how First Physiognomy proceeds to Climax Physiognomy (Figure 10.7).[24] Honda had studied in Munich from 1888 under

Gayer, who had insisted on 'natural regeneration' to form mixed forests and called for society to go 'back to nature' in 1878. Mayr was one of the foreign advisors hired by the Japanese government. Their ecological view ran counter to the economically efficient monocultural planting of coniferous trees in favour in German society at that time.[25] Honda and Hongo realized that the way to grow a 'natural' forest, that is, a broadleaved evergreen grove, depended on an ecological process which started from a mixed forest using other scenic trees over the long term.

Tree donation was a rather successful measure, carried out from the first planning stage in the fiscal year 1914. Eventually, more than 95,000 donated trees were collected, 89.6% of the total number of planted trees.[26] People were sincerely delighted at this dedication and cooperated willingly. Some tried to send precious ancestral trees, local rare trees or unusual foreign trees to show their dedication. However, the intention of the grove creation plan was to form the eternal forest as an integrated whole. A list of 91 desired species was attached to a request for donated trees to make the initial mixed forest. It said that 'the donated tree should be conditioned to adapt to the natural forest of the divine precinct and a garden tree would be inadequate'.[27] These donated trees were obviously not for a beautiful garden but for a divine thick forest.[28] The ultimate purpose was to form an eternal evergreen grove which could represent the people's allegiance and admiration.

What was left after this period?

Recent studies have drawn attention to the vegetation of the shrine grove from a practical point of view. The shrine groves which were planted around Kyoto at the beginning of the twentieth century were mainly composed of sparse red pine trees with deciduous trees such as cherries and maples, to provide colour.[29] But Seiji Tsukamoto, who succeeded Tomoichi Inoue as chief of the Meiji Shrine Construction Bureau (1915–21), insisted that there should be a clear distinction between the shrine and public parks, even though most people then could not differentiate between them. In 1942 he wrote that the shrine should be designed as both tranquil and strained, as if there was a divine presence in the evergreen trees. A park, by contrast, was just an enjoyable place to stroll at leisure. He insisted that shrine groves and public parks were incompatible.[30] His opinion reflects the aesthetics of the Meiji Shrine, and shows how bureaucrats were influenced by its construction.

The Meiji Shrine was designed with its two parts: its Internal Precinct was a grove, and the External Precinct corresponded to a public park.[31] The use of 'dense forest' was supported by the shrine administration officials, and the provision of 'open enjoyable space' was supported by the town planning and landscape experts.[32] In this re-evaluation of the desired appearance and maintenance strategies for the local shrine groves, a new visual image of a broadleaved eternal forest became the model which was established by the forest engineers of the Meiji Shrine construction.

The Internal Precinct of untouchable evergreen forest was connected to preservation measures based on the Preservation Law of 1919.[33] This law was another consequence of Minakata's opposition to the shrine mergers. The visual image of this traditional shrine grove provided an ideal appearance for their preservation (Figure 10.8).

The government also sought to establish another expression of local autonomy, a park. Inoue explained it in his 'Essence of Autonomy'[34] by saying that 'the park policy' was not only a measure for health and hygiene, 'the lungs of the city', but it was a social education measure. Notably, he claimed that the community education expected of the shrine was also part of a park's function. Kosuke Tomeoka, his subordinate, had already conceived this idea in 1906, the same year the Shrine Merger policy started. Tomeoka noted that this had come from the influence of Inoue's article, 'European Parks', but he also wrote that a park should be a place where people share public space to learn 'public morality' and to nurture their minds, as well as to cure their fatigue.[35] As the dust settled on the Shrine Merger, the idea of a place to nurture citizens' minds was transferred to the public parks policy.

樹木（上）

農學博士 川村清一

淳仁天皇淡路御陵全景

Figure 10.8 An article 'Imperial Tomb and Its Trees' in *The Historical Spot, Scenic Beauty and Natural Monument Preservation* 5:10 (1920), pp. 111–12, discussed how a dense grove ought to be maintained in order to preserve the setting of the shrine and divine mausoleum.

Kazumi Iinuma, who was manager of the City Planning Section (1931–4) and inheritor of the bureaucratic chain of thought descended from Mizuno, demanded 'a PARK per town or village' in 1933.[36] Bureaucrats focused their attention on city and regional planning as the main vehicle for local autonomy, while the 'eternal' grove became a preservation target as a 'scenic zone' or as a 'green zone' within the planning system.

Conclusion

Although it might be believed that the shrine grove has kept its natural vegetation since the primeval era, and that it has always looked like the modern shrine grove, dense and tufted broadleaved evergreen woods, the physiognomy of the grove actually changed dramatically in the modern epoch. This change was a consequence of the struggle to fashion the modern Japanese nation, and the subsequent interaction between old Japanese cultural perceptions and those imported from Europe under the Meji regime and its successors.

In the early twentieth century the local development was an urgent problem. The government bureaucracy tried to establish a system of local autonomy which could take priority over the national construction projects which were affected by unstable party politics. A focus of this strategy was the shrine. Officials saw them as being at the heart of a traditional community, at the centre of reverence for the home village, and they proposed establishing the new local autonomous system with the shrine as a traditional metaphysical element, a node in a national network. However, the building of this autonomous system was not completed as the Home Ministry planned; replacing the shrines did not work as a way of reconstructing effective local autonomous government, even though the strategy had a relatively rational underlying ideology, aiming for measurable, bureaucratic performance, 'bureaucratism'.[37]

Because of the opposition to this strategy, the value of the existing local shrine groves acquired a purpose. The shrine groves became a symbol of shared history and landscape which strongly connected to villagers' daily lives and their reverence for their ancestors. This idea of an ideal shrine grove evolved into 'the broadleaved evergreen grove', adopting the appearance of the Meiji Shrine grove which had only been created as a practical response to the needs of the site selected for it. There had been technical reasons to choose broadleaved instead of coniferous trees, rooted in the German foresters' knowledge of natural vegetation and the notion of natural regeneration. The preservation of the shrine precinct as a divine place, and its consequential neglect,[38] was directly influenced by the completion of the Meiji Shrine and the aesthetic and practical decisions made in its construction.

The public park became the place which served local autonomy, particularly after the measure to merge shrines was shown to be an imperfect solution. Bureaucrats focused their attention on city and regional planning in which the public parks were significant elements. The 'eternal' grove became a preservation target, part of a scenic or green zone.

The Shrine Bureau and shrine systems connected to an Imperial faith were dismantled and reformed by the General Headquarters of the Allied Powers after World War II, and since then the relation between shrine and people has been weakened. The shrines have become one amongst many forms of religious activity, and they have only kept their connection with people through annual events like the New Year or Harvest Festival. The shrine groves have been less controlled, without active maintenance, and their physiognomy has tended to slip towards their potential natural vegetation, broadleaved evergreen forests, especially in the middle and south-east of Japan. The naturalism and the modern image of the shrine grove have helped this process. Nowadays, *Chinju no Mori*, with dense and tufted forest, is described as a 'typical Japanese landscape' and as 'evidence of local patriotism',[39] and has been refocused as a precious common 'natural' place for citizens, who do not realize that its character is little more than a century old.

Notes

1 Japanese oaks include several evergreen species: *Quercus phillyraeoides* A. Gray, *Quercus acuta* Thunberg, *Quercus glauca* Thunberg (*Cyclobalanopsis glauca* Oerstedt), *Quercus gilva* Blume (*Cyclobalanopsis gilva* Oerstedt).
2 Makatu Numata, Akira Miyawaki and D. Ito, 'Natural and semi-natural vegetation in Japan', *Blumea* 20:2 (1972), pp. 435–96.
3 Minoru Sonoda and Masanobu Hashimoto, *Dictionary of Shinto History* (Tokyo: Yoshikawa-Koubundou, 2004), pp. 677–78.
4 Junichi Ogura, *History of Radically Changed Vegetation of Japan* (Tokyo: Kokon-Shoin, 2012), pp. 256–337.
5 *Dictionary of Shinto* (Tokyo: Kobundou, 1999), p. 91.
6 Tetsuo Najita, *Japan: The Intellectual Foundations of Modern Japanese Politics*, (Chicago: University of Chicago Press, 1974); Junji Banno, *The National Vision of Modern Japan 1871–1936* (Tokyo: Iwanami Shoten, 2009), pp. 2–6.
7 Yorio Fujimoto, *Modern History of Shinto and Welfare Works* (Tokyo: Kobundo, 2009), pp. 11–52.
8 Tomoichi Inoue, *Essence of Autonomy* (Tokyo: Hakubunkan, 1909), p. 34.
9 The Shrine Bureau was established in the Home Ministry in 1900. When Mizuno became chief of the Shrine Bureau in 1904, it tried to institutionalize the municipal offerings to the prefectural and village shrines, to consolidate the small, decrepit shrines. However, the Provincial Bureau, led by Tomoichi Inoue, opposed this because provincial municipalities had budget shortages. After some bargaining, the Provincial Bureau allowed municipalities to pay the shrines' ceremonial offerings. The Provincial Bureau's trade-off condition was that the Shrine Bureau should cooperate to bring autonomy to the provincial municipalities. This idea seemed beneficial and was one that Mizuno pushed forward eagerly. See Kiyomi Morioka, *Local Shrine and National Control in the Modern Period* (Tokyo: Yoshikawa-Koubundo, 1987), pp. 81–107.
10 Rentaro Mizuno, 'Shrines should be the centre of the municipality' (1908), in Katsuya Hidezo (ed.), *The Application of Autonomy System and People* (Tokyo: Jitsugyo no Nihon Sha, 1912), pp. 109–26.
11 Wilbur M. Fridell, *Japanese Shrine Mergers 1906–12* (Tokyo: Sophia University, 1970).

12 The Statistics Bureau of the Cabinet, *Statistical Yearbook of the Empire of Japan*, 1912 and 1924.

13 Kumagusu Minakata, 'The letter sent to Matsumura Jinzou' (1911), in Shinobu Iwamura, Yoshitaka Irie, and Seizo Okamoto (eds), *Minakata Kumagusu Zenshu* (Tokyo: Heibonsya, 1971), p. 493.

14 Kumagusu Minakata, 'The oppositional statement against the shrine merger' (1912), in *Minakata Kumagusu Zenshu*, pp. 566–94. The original article was published in *Japan and Japanese* 580, 581, 583, 584 (1912).

15 Junichi Ogura, *History of Radically Changed Vegetation of Japan* (Tokyo: Kokon-Shoin, 2012), pp. 329–35.

16 Jingi Council Academic Affairs Bureau Research Division, *Stories of the Epoch of the Shrine Bureau* (Tokyo: Jingi Council, 1942), p. 88.

17 Kazunori Sato, *Study on the Theory of the Meiji Emperor's Virtue* (Tokyo: Kokusho Kanko Kai, 2010).

18 For details of the External Precinct, see Yoshiko Imaizumi, 'The Making of a Mnemonic Space: Meiji Shrine Memorial Art Gallery 1912–1956', *Japan Review* 23 (2011), pp. 143–76.

19 The Bureau of the Meiji Shrine Construction, *Commemorative Issue of the Meiji Shrine* (Tokyo: The Bureau of the Meiji Shrine Construction, 1923), p. 1–23.

20 Keiji Uehara, *The Forest That People Created* (Tokyo: Tokyo University of Agriculture Department of Landscape, 1971), pp. 10–25.

21 Seiroku Honda had learnt directly from Karl Gayer who insisted on natural regeneration based on the natural ecology. See: Yoshiko Imaizumi, *The Meiji Jingu: The Great Project to Create a Tradition* (Tokyo: Shincho Sensho, 2013), pp. 124–29.

22 Yoshiko Imaizumi, 'The Making of a Mnemonic Space: Meiji Shrine Memorial Art Gallery 1912–1956', pp. 143–76.

23 The argument between Seiroku Honda and the Prime Minister Shigenobu Okuma is interesting. Okuma insisted on an avenue of great cedars referring to Nikko. Honda and his team struggled with scientific data collection to show the maladaptation of cedars at the site. See Uehara, *The Forest That People Created*, pp. 20–5.

24 Takatoku Hongo, 'The Grove Plan of the Meiji Shrine Site', 1921, in *Meiji Jingu Library* 13 (Tokyo: Kokushokankokai, 2004), pp. 447–717.

25 Karl Hasel, *Ein Grundriß für Studium und Praxis*, trans. Mitsuaki Yamagata (Tokyo: Tukiji Shokan, 1996).

26 Although the method of donating trees was planned by Zentaro Kawase and Seiroku Honda's team, the idea to donate trees was proposed by Tomoichi Inoue. It was a way of converting the intangible property of people's sincerity into tangible memorabilia. It was in the context of Inoue's Local Improvement Movement by using the shrine's mental centripetal function, which was necessary for local autonomy, and by nurturing cooperation and the national conscience. See Sato, *Study on the Theory of the Meiji Emperor's Virtue,*, pp. 235–41.

27 Cabinet Official Gazette Bureau, *Official Gazette*, 6 March 1915.

28 Uehara, who became an official of the Meiji Shrine Construction Bureau, admitted later that he rejected the offers of precious garden trees with a letter stating the original purpose carefully. (Uehara, *The Forest that People Created*, pp. 61–7)

29 Setsuko Nakajima, 'Forestry and Landscape in Kyoto in the First Half of the Meiji Era', Architecture Institute of Japan, *Journal of Architecture Planning and Environmental Engineering* 481:3 (1996), pp. 213–22.

30 Seiji Tsukamoto, 'Notes for the Shrine Precincts', in *Stories of the Epoch of the Shrine Bureau* (Tokyo: Jingi Council Academic Affairs Bureau Research Division, 1942), pp. 121–26.

31 Yasuto Noma, *Ideal Gardens and Parks* (Tokyo: Nihonhyoronsya, 1923), pp. 250–55.

32 Setsuko Nakajima, 'The Creation of "Shin-En (Shrine Garden)" in Kyoto in the Modern Period', Architecture Institute of Japan, *Journal of Architecture Planning and Environmental Engineering* 493:3 (1997), pp. 237–43.

33 Three Natural Monument categories were designated: animals, plants and geological minerals. Top of the list in the plant category was 'shrine groves': See: Nishimura Yukio, *Urban Conservation Planning* (Tokyo: University of Tokyo Press, 2004), pp. 53–6.

34 Inoue, *Essence of Autonomy*, pp. 170–73.

35 Kosuke Tomeoka, 'City and Parks' and Kosuke Tomeoka, 'The Ethics of the Park' (both originally published in 1906) in *Tomeoka Kosuke's Collected Works*, vol. 2, (Tokyo: Doshisha University Institute for Study of Humanities and Social Sciences, 1979), pp. 238–43.

36 Kazumi Iinuma, *The Theory of Regional Planning* (Tokyo: Ryosho Fukyukai, 1933), pp. 109–12.

37 Najita, *Japan: The Intellectual Foundations of Modern Japanese Politics*, p. 2.

38 Ogura, *History of Radically Changed Vegetation of Japan*, pp. 330–32.

39 Sonoda and Hashimoto, *Dictionary of Shinto History*, pp. 677–78.

11 Nature mystification and the example of the 'heroes' groves in early twentieth-century Germany

Gert Groening

In December 1914, four months after the outbreak of World War I, Willy Lange (1864–1941),[1] a landscape architect, then Royal Horticulture Director and professor at the Royal College for Gardeners' Training (*Königliche Gärtnerlehranstalt*) in Berlin-Dahlem, Germany, suggested creating *Heldenhaine*, 'heroes' groves' and 'holy groves' for soldiers who had died for the German Reich.[2] Adjacent to most cities in Germany, small and large, heroes' groves were to be established for the commemoration of fallen soldiers. The design concept for these groves relied on only a few elements. Paths would lead to an enclosed inner area with a linden tree in the middle surrounded by a ring of oaks (Figures 11.1 and 11.2). Stones inscribed with names of fallen soldiers were to be placed within the ring of oak trees. A wall and a ditch would surround the grove and designate its outer border.

Lange's design proposal differed notably from those of earlier well-known sculptural and architectural memorials in Germany, like that of 1899 for the Prussian first guard regiment of foot, which showed a bronze sculpture on a large rock pedestal of an armoured Saint Michael angel leaning on his sword. Traditionally war memorial design in Germany followed Christian and classic-ancient iconographies. Some used martial signs to celebrate war, even in death.[3] Lange's concept, however, proposed to do without traditional iconography and obvious references to warfare. From a social point of view Lange's concept appeared egalitarian. For every killed soldier one oak tree was to be planted. The grove was supposed to serve as both a personal memorial and as a stage for various ceremonies. During wartime, the linden tree in the grove would symbolize the hope for peace.

Lange took his design ideas from sacral sources, wanting heroes' groves to become appreciated as holy places, but on closer observation his concept pushed accepted traditions too far. His idea expressed nationalistic-monarchistic thinking and displayed a racist world view. It also stood for aggressive militarism. Only a few heroes' groves were implemented, one at Ludwigslust in Mecklenburg-Vorpommern and another at Lehmkuhlen-Hohenhütten in Schleswig-Holstein, Germany, where an oak was planted for each fallen person from the estate.[4]

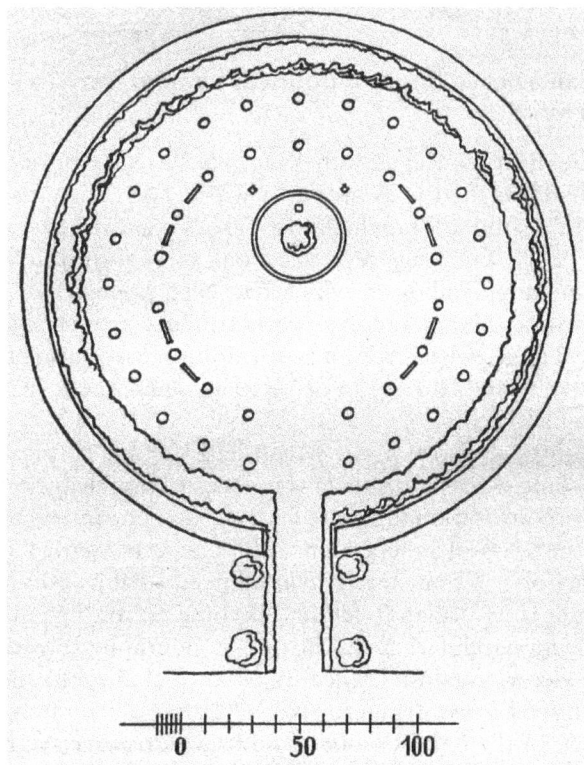

Figure 11.1 Hero grove: Linden and 40 oaks with tree mound and ditch, design Willy Lange (Willy Lange (ed.), *Deutsche Heldenhaine* (Leipzig, 1915), p. 34).

Figure 11.2 Oak Grove at a round-shaped mountain with spiral walk, circular clearing and memorial hall, Hohenstaufen near Göppingen, South Germany, design Felix Genzmer and Emil Fader. Hohenstaufen village and its church are to the right (Willy Lange (ed.), *Deutsche Heldenhaine* (Leipzig, 1915), p. 45).

Heroic oaks and peace lindens: political iconography to serve monarchy

The holy groves' basic concept was to plant oaks as a memorial for German soldiers who had lost their lives in World War I. For Lange the oaks served as a kind of 'life memorial' in which 'the heroes continue to live for us and the centuries'.[5] Early on Lange revived an oak tree motif well known in literature, painting and symbolism which had been popularized in Germany by various patriotic, liberal and even revolutionary groups since the eighteenth century.[6] Lange twisted the ancient tradition associating the oak with strength, constancy and courage in order to associate these virtues with the groves for German heroes.

Lange used the oak grove motif which had become so meaningful during the Napoleonic wars. In those days it was still not clear how the nation should become constitutionalized, federal or unified, imperial or republican.[7] During the wars of independence the oak was seen as a symbol for an abstract idea of a nation-state which allowed various political concepts to develop. After the German Reich had become established in 1871 it was usurped for the monarchistic idea. The oak-iconography spread everywhere following the victory against France in 1870, including to the plantation of so-called *Kaiser-Eichen*, (emperor-oaks).[8] Lange deliberately ignored the former meaning of the oak as a liberal and revolutionary symbol. For him the fight for German freedom and unification belonged to a monarchistic tradition.

Lange viewed the freedom fight against Napoleon as a political precondition to establish a German empire: 'When peace came in 1871, one planted emperor-oaks in the sense of the acquired unity and the acquired emperordom'.[9] Lange also compared Germany's World War I struggle to that of the German states against Napoleon: 'We all know: the oak has become a German people's-tree since 1813, the tree of freedom from foreign suppression; and it is ultimately that which was fought for again in 1914'.[10] In order to underline the monarchistic idea Lange accented it in a characteristic way. A linden tree should form the centre point of the oak grove, to represent Emperor Wilhelm II (1859–1941).[11] So, Lange's heroes' groves deliberately propagated monarchism.

Lange envisioned another far-reaching use for the oak, turning it into a meaningful symbol of unbroken national tradition to recapture a Teutonic past which began in Germanic prehistory and ended with Germany's international political claims during World War I.[12] 'Old-Germania as origin – 1813 Germany as longing – 1870/71 Germany as fulfilling in itself – 1914 Germany as holding world recognition'.[13] Lange wanted history to be a cyclic process, so he construed an undue historical continuity for German history.

A comparable historical pretence developed when the 'Walhalla' building was erected near Regensburg, Bavaria between 1830 and 1842 by architect

Leo von Klenze (1784–1864). *Walhalla* was the court of Woden, the high-est god in Germanic mythology. He was widely known as a god of war, and *Walhalla* was his hall for slain heroes. This first 'all-German national memorial' had been conceived of under the Bavarian crown Prince Ludwig (1786–1868) in 1807, during the Napoleonic War. Its interior contained the busts of great Germans surrounded by a frieze which displayed Germanic prehistory, the life of the Germanic people until Christianization and presen-tations of Germanic mythology.[14] Following Eric Hobsbawm, this strategy of political legitimization may be described as the invention of tradition.[15]

For Lange the political reference to ancient Germania opened up a reli-gious perspective. He believed he touched upon the tradition of sacred or holy groves as they were known to the Teutonic people[16] and referred to the Roman historian Tacitus: 'Since Tacitus has described Germanic worship for gods in holy groves, the idea of the sublime has always been connected to the oak as the venerable forest tree. In this idea the oak became the holy tree of the Germans'.[17] In his *Germania* Tacitus had pointed out several considerable differences in god-veneration cults between Germanic tribes and Romans: Germanic tribes worshipped their gods in holy, invulnerable groves.[18]

Lange's heroes' groves were also supposed to possess and enjoy a sacred protection: 'No flower may be picked. No tree trunk may be damaged or carved with letters and signs. The grove must be holy also in the sense of invulnerability'.[19] This led to the encompassing of heroes' groves by a wall and a ditch: 'Protection is needed. . . to create the feeling of seclusion, security in its interior, also for the impression of separation from every-thing general; because this is what everything specific, holy, and dedicated demands'.[20] Heroes' groves should also be sacred as war memorials and as symbols for the patriotic-monarchistic principle, and for that a special Ger-manising iconography was developed.

Old Germanic monuments as sacred proof of Germanic ideology: circle and *Walburg*, battle fortification in heroes' groves

Lange's first text of December 1914, 'Heroes' Oaks and Peace Lindens', called for a new type of memorial and compared it to prominent landscape features of Germanic prehistory. He linked heroes' groves to buildings of contemporary German history, the Bismarck towers. All over Germany these towers had been erected in large numbers in late nineteenth and early twentieth centuries in honour of Prince Otto von Bismarck (1815–1898) who had died in 1898.[21] Frequently Bismarck towers would offer a view over the area. 'When the German people built their Bismarck towers', Lange wrote, 'it found the architectural expression of its tribal union in a uniform landmark in all German states'.[22] In this context Lange referred to a num-ber of features characteristic of early Germanic history which he believed had comparably coined local areas: 'Stone settings, sun circles, altars, hero marks, and rune-stones were erected, high-reaching, where Germanic people migrated, were victorious, and settled: a mark proper and a tribal sign imposed stone marks upon the landscape – as did Bismarck Towers to the newly united shires of Germany'. Lange saw Germany's development 'between [the Celtic] menhir and the Bismarck tower'. According to Lange they had in common their verticality: 'Upright, directed towards light, has always been what Germanic spirit manner has created'. Significantly, here Lange mentions the culture historian Willy Pastor (1867–1933) who 'in his numerous treatises "from the North" has acquainted us with this spirit'.[23]

Pastor played an essential role for Lange's heroes' groves concept. Pastor had introduced as characteristic features of the holy groves the ring structure of the mound and the ditch (Figure 11.1).[24] In 1915 he wrote about 'the meaning of the circle in the heroes' groves': 'It is no accident and no artistic gimmick if it is the ring which dominates the entire site. Here also old and oldest awaken again from Germanic pre-historic times. The ring or several concentric rings served as sun and sun course symbols'.[25] The ring motif was connected to sun worship in prehistoric Germanic times, and Pastor pointed to the sun temple at Stonehenge in England as an example.[26] Like Lange, Pastor used the concept to politically legitimize the German empire. Pastor categorically claimed megaliths like Stonehenge were directly related to Germanic culture. This is surprising since Pastor himself, as well as most of his contemporaries, were well aware of the almost ubiquitous existence of megalith cultures, and especially of stone circles.[27] In a number of publications Pastor outlined worldwide megalith introduction by a primeval Germanic race.[28] His 1910 book, *Altgermanische Monumentalkunst* (Ancient Germanic Monumental Art), elaborated on an alleged Germanic cultural transfer. First the Germanic race was said to have reached Northwest Europe and the British Isles. Via the British Isles the Germanic people had then gone to other parts of the world and always left stone witnesses of their culture.[29] Pastor found 'one can not deny that megalith culture once originated from the Germanic North'.[30]

Pastor based his claims of a Germanic empire upon Gobineau's and Chamberlain's views of a leading Aryan and especially a Germanic race. Count Gobineau (1816–1882)[31] had tried to explain the origins of peoples in the history of mankind. He attempted to conceive an idea about the historical origin and the position of peoples within the history of mankind and civilization.[32] His basic hypothesis was that ultimately all people derived from one common (Adamite) tribe which had spread all over the globe.[33] This tribe later split into three main races, the black, the yellow and the white. The white race developed human civilization.[34] It was an 'Aryan race' which had spread across the entire earth in early times.[35] Germanophile Englishman Houston Stewart Chamberlain's book *Die Grundlagen des neunzehnten Jahrhunderts* (The foundations of the nineteenth century) followed such ideas.[36] However, this Englishman and self-elected-German corrected Gobineau's pessimistic view of the imminent end of the Germanic race and added an anti-semitic tendency to it. Like Gobineau, Chamberlain believed in the leading cultural and racial role of the Germanic people.[37] Pastor's talk of a Germanic noble-race is evidence for the racially founded, supposed cultural superiority of the Germanic people since the beginning of the Neolithic age.[38]

Mount Hohenstaufen near the south German city of Göppingen served as an almost perfect example for such a *Walburg* (Figure 11.2) because it was not an artificial mountain, but real. The shape of the mountain is a result of regressive erosion along prehistoric shores of the Suebian Sea. The top of the mountain still carried the ruins of an eleventh-century castle, *Stammburg*, of the Hohenstaufen dynasty. More than an artificial mound could ever achieve, this mountain provided almost ideal conditions for a hero grove (Figure 11.3). For Lange it was an outstanding example:

> Where hills allow the establishment of a hero grove there it will have its most beautiful and strongest effect and will offer the richest chances for great art works of building. To name just one location as an example of which there are many in Germany where this basic shape is given, the Hohenstaufen shall be mentioned. Here at the same time the deepest relation to Germany's greatness, grown out of long history, would be gained.[39]

In Lange's book, *Deutsche Heldenhaine*, architects Felix Genzmer (1856–1929) and Emil Fader (1885–?) provided a visualization of the *Hohenstaufen* as an oak grove with a spiral path, a circular clearing and a memorial hall.[40]

Lange followed Pastor's view of *Walburgen* as former sun-sanctuaries from Germanic primeval times which symbolized the course of the sun. 'And if nothing else, the *Walburg* is a sculptural representation of the celestial castle, that powerful world-mountain, which the liberated sun regains every year in springtime'.[41] The *Walburg* motif would not only revive a built monument but also a cultish dimension. Lange considered it negligible if a Christian or a heathen meaning was attributed to the site: 'The imitation of the sun-course which was believed to be spiral became the symbol for Christ's path of passion with his death of salvation at the end. . . who symbolically followed this path, found salvation in the sky, – who once went the sun-path, found the way to *Walhall*'.[42]

This allowed fallen soldiers to be symbolized by oaks, either as Christian martyrs or as heathen heroes. In both instances death by war became idealized as a sacrifice to increase the readiness for sacrifices. Whether one decided on the more costly design of a *Walburg* or of a ring for a hero grove was insignificant for an evaluation of the ideological statement. The motifs of the *Walburg* and the ring served as vehicles for a racial global view. Ultimately the goal was the re-creation of a Germanic, a German, domination of the world, as Pastor indicated.[43] This view shaped Lange's hero grove

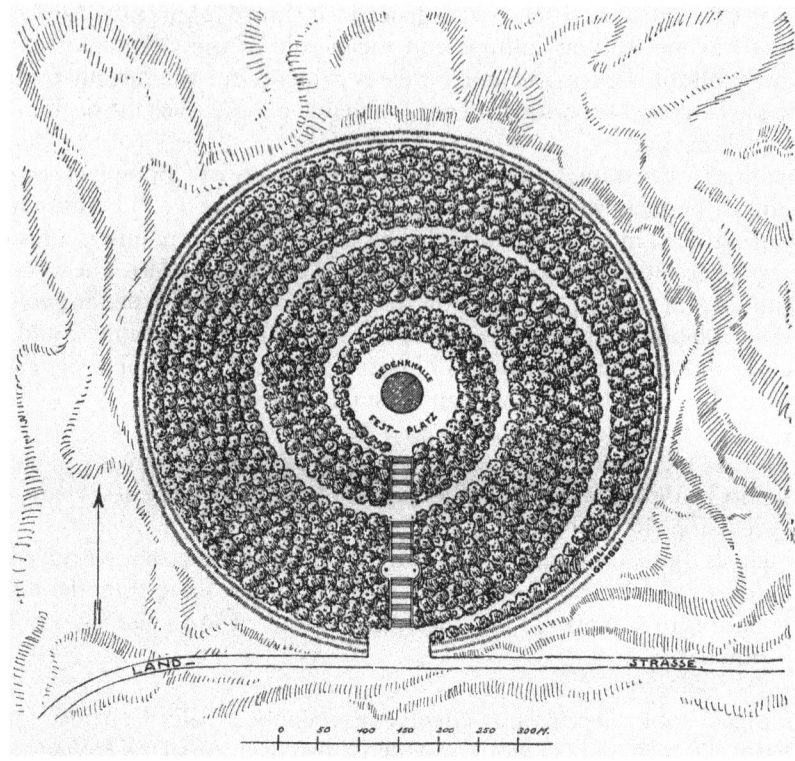

Figure 11.3 Oak Grove on mountain top with spiral path and stair walk. Memorial hall on top with fairground, design Felix Genzmer and Emil Fader (Willy Lange (ed.), *Deutsche Heldenhaine* (Leipzig, 1915), p. 44 (compare Figure 11.2)).

concept in which World War I was a racial conflict: 'Everything which joins with the Germanic racial community will win: both the German idea and the German God want this'.[44]

Eventually Lange saw victory in the war as offering a kind of relief for the world: 'What is the greatest? . . . One in all: the idea of the Germandom which wants to save the world by itself and by the sacrifice of every single person!'[45] So the concept of the heroes' groves points far beyond a memorial. It would become a political instrument for the hope for Germany to become a great political power.

Heroes' groves as means of military education for the youth

Lange's hero grove concept had a clear interest in militaristic youth education. By 1914 Lange had already associated the notions of militarism and heroes' groves: 'The larger the grove of a fallen community is, the greater their honour and the honour of the fallen members of the community. It also would be a memorial for victorious militarism if the memorial-oaks would stand in rank and file'.[46] Thus, one reason for positioning the oaks in a ring is to give the quasi-military formation the character of a memorial. The heroes' groves should promote military youth education and should support physical power as well as mental readiness for fighting.

Johannes Speck, a Berlin high school teacher,[47] contributed a text, 'Heroes' Groves and Youth Care' to Lange's *Deutsche Heldenhaine* publication. Speck pointed out how meaningful a state memorial cult for the soldiers fallen during World War I would be.[48] He also outlined the exemplary meaning of such memorials for youth. Following a few thoughts about the diversity of the emerging early twentieth-century youth movement[49] which made him question where this youth was heading, Speck found that the German youth of World War I 'delivered a youth image hardly conceivably more beautiful and impressive. At one stroke the threatened fatherland filled young men of all social ranks with the same thoughts and ideals, they all show the same readiness to give their lives for the same goods'.[50] However, Speck questioned if such spirit would endure after the war: 'Will social relations suitable to undermine youth idealism not begin their work again after the end of the war?' For Speck this justified a double approach. All those who had already joined forces to sustain youth power for the German people before World War I should keep this in mind for peace times, in order to carry the prevailing spirit of youth idealism into the new social relations of peace.

'If, however, signs and symbols are able to bring these forces together and guide them, then', Speck believed, 'to a high degree it is the heroes' groves suggested by Willy Lange'. Speck continued: 'Even the thought that oaks lead us out from the narrowness of the city makes our soul feel alleviated and liberated. Groves, however, at the same time lead us to the eternally young and juvenating ground of nature within the heart of humans which our warriors revealed through their deaths for the fatherland'. Speck likened the effects of the ancient Roman saga about Philemon's and Baucis' metamorphosis into an oak and a linden tree to the impressions heroes' oaks would evoke.

Heroes' groves would educate the youth in a military fashion. 'Like games and gymnastics, the military exercise of our youth. . . will take place near the hero grove, the most suitable place for them'. The heroes' groves were considered a perfect locale for physical and mental military training, and landscape architects would suggest *Wehrgärten*, (Defence Gardens) for youth parks (Figure 11.4). Heroes' groves would also serve as locations for games and singing competitions. Ultimately Speck envisioned heroes' groves similar to heroes' memorials in 'old-Germanic time', in Antiquity, and in the Middle Ages, as *foci* and centre points of national and religious life: 'Our youth and people's games, our gymnastics, our sports and soldier-like

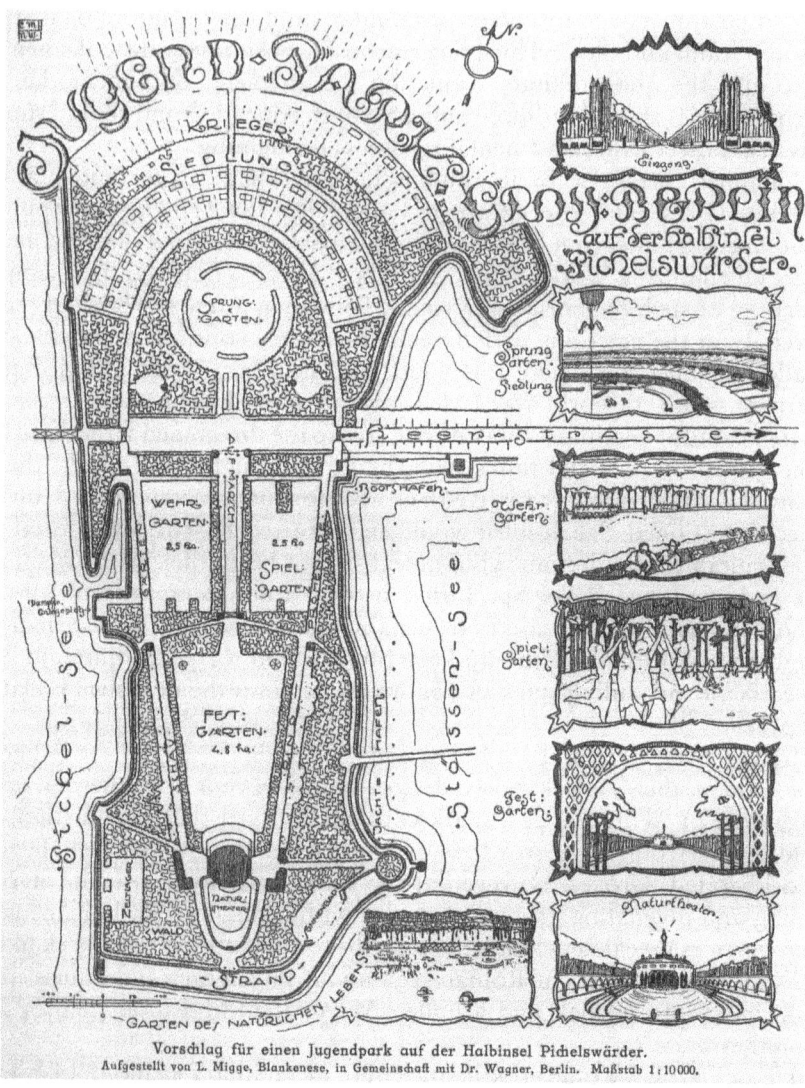

Figure 11.4 Wehrgärten im Jugendpark, defence garden in a youth park, design Leberecht Migge and Martin Wagner. Within the park the defence garden is located in the middle to the left. It is represented by sketch three from above to the right, where boys throw grenades from a trench (Leberecht Migge, 'Jugendparks als Kriegerdank', *Die Gartenkunst*, 29: 9 (1916), pp. 120–4 (121)).

exercises will, if located at the foot of hero-oaks, get something from the dedication of the Greek games which had received their high meaning from the fact that they were held near their temples and had been brought in close relationship to their gods' worship'.

Preparation for war service thus turned into a sacred act. Death as a consequence of war appeared less threatening if interpreted as sacred sacrifice for higher goals. In a characteristic way Speck's concept combined military, sports and play activities within the bounds of youth education. Military and nationalistic interests thus shaped youth service. What may appear unusual in early twenty-first-century Europe was common during the German empire with its then general tendency towards militarization.

Specifically this interest dates back to successful attempts to influence education politically in Prussia at the end of the nineteenth century. In 1890 Gustav von Gossler (1838–1902), then Prussian minister for spiritual, educational and medical matters, organized a school conference on 'Questions of Higher Education' in Berlin.[51] Participants included politicians, administrators, university professors and officers from the war ministry. Because some deficits had been detected, contributions to the conference addressed the meaning of national, military and, in relation to it, educational goals for physical training in schools.

Significant at this conference was a programmatic speech delivered by Emperor Wilhelm II (1859–1941, emperor 1888–1918) himself. He stressed that 'our growing youth. . . must be suitably educated. . . in relation to the global position of the Fatherland'.[52] Instead of humanistic education the emperor wanted more patriotic history. He complained that pupils 'lacked above all the national basis',[53] and disagreed with what he considered an excess of intellectual challenge in the classroom. Instead he wished to improve physical abilities through sports activities, so he pointed to military education especially: 'Consider how a rising generation for state defence grows up. I seek soldiers. We want to have a strong generation'.[54]

A number of presentations at the school conference explicitly addressed questions of national and military education. A Dr Goering pointed to the meaning of history lessons. One should prefer 'books which portray the life of the German emperor', and in practical education Goering demanded a stronger consideration of 'gymnastics, youth games and singing'.[55] Fleck, who held the military rank of major, addressed the meaning school education had for the 'fighting-power of our people and thus for the position of power for our fatherland'.[56] With respect to war conditions he demanded the communication of moral properties, that is, a kind of mental readiness for fighting: 'We should enforce the moral factors in man. . . and strengthen them with the ultimate goal that a soldier in the case of war, in the hours of danger is not found wanting but consciously and readily follows us to death for emperor and empire'.[57]

Another participant in the school conference made clear that besides physical also spiritual fighting power needed to be faced.[58] This school conference of 1890 had created essential preconditions for the 'total politicization of school' for national and military purposes. It called for the establishment of a 'central committee for the promotion of people and youth games in Germany' founded one year later.[59] The games included gymnastics as a special target to train for future service for the fatherland.[60] This was in line with Wilhelm II's speech before the school conference in 1890.

It was the goal of the 'central committee for national and youth games' to oppose youth's lack of military usefulness and the topics of the conferences of the 'central committee' reflect this interest. The second school conference in 1893 discussed the use of youth and people's games for the army. At the third school conference in Königsberg in 1899, the Quedlinburg school director Hermann Lorenz (1860–1945) asked in his keynote speech, 'What demands has the army service for the moral and physical properties of youth, and how can youth education which serves the national fighting-power create the preconditions for that?'[61] Lorenz then discussed the value of 'bourgeois virtues' such as fear of God, loyalty to the emperor, love of the fatherland, national feeling of honour, readiness for sacrifice with respect to a not too distant war and obedience.[62] Especially he elaborated upon the techniques of pre-military youth education. The exercises and games he suggested should support the development of power, perseverance and becoming inured to fatigue. Exercises for perseverance and running should prepare for marching and 'quick advance'. He described the people's and competitive games with respect to properties 'which a man needs in a war'. In 1899 a 'committee for the promotion of military power through education' was founded. It included military people, politicians, medical professionals and pedagogues.[63] According to Lemmermann the guidelines which this committee had formulated in 1904 'provided for a comprehensive system of influence in order to guarantee with children and youth physical as well as psychical dispositions for optimal development of war-abilities'.[64]

The introductory remarks for the guidelines read: 'Every teacher who cares for his people and for the future of the fatherland must for his part, with all his power, see to it that the state has at its disposal courageous strong young blood for the army'.[65] Besides the compilation of the [report of the] 'committee for the promotion of military power through education', a number of other publications appeared around 1900 which also referred to 'military power through education'. In 1899 Hermann Lorenz published a book with just that title.[66]

First, though, it was up to more singular initiatives to implement such a programme. In several German cities youth brigades practised war games in 1896.[67] Officers of a Munich garrison founded the organization *Wehrkraft* ('military power') in 1909. By 1913 it included 62 local groups and 9,000 young members. These youths had been trained in target shooting as well as in other military skills.[68] In 1911 the German Pathfinder Union was founded. Following the example of Robert Baden-Powell's (1857–1941) *Scouting for Boys* (1908), Alexander Lion (1870–1962) and Maximilian Bayer (1872–1917) published the manual *Pfadfinderbuch* in 1909. It was an adaptation of Baden-Powell's book which contained an army-oriented programme for exercises.

In Prussia the various associations were systematically brought together when the *Jugendpflegeerlass* (youth-care-decree) was issued in 1911. On community, county and regional levels so-called *Jugendpflegeausschüsse* (youth care committees) were established, aiming for a 'patriotic education'. The *Jungdeutschland-Bund* (Young Germany Association), established in 1911, systematically unified the various youth organizations. By 1914 it counted about 750,000 members and co-operated closely with the military and employed many military instructors. The hiking sections of the youth association were offered army barracks, exercise fields, gymnastic halls and training grounds.[69] When the *Jungdeutschland-Bund* was founded, members of the *Wandervogel* (Migrating Bird organization) pointed out that they had long been active in military education of the youth.[70]

Long before World War I youth education had been systematically influenced by and geared towards military purposes. Speck's concept of military education formed part of this tradition.[71] His activities as well as those of the 'central committee' of 1891 and of the 'committee for the promotion of military power through education' of 1899 expressed a common tendency towards militarization by imposing military principles upon other aspects of life. Ultimately this permeated bourgeois life with military principles.[72] This socio-political dimension of the notion of militarization is clearly visible in Speck's contribution. He not only connected youth education to military purposes, but defined the army itself as an 'education institution' of the people: 'And in spite of all that has been said against it occasionally it is true nevertheless that our army is an educational institution in such a way that it develops commonly valuable virtues'.[73]

The army as an 'educational institution' of the people was a concept developed by the military. The Prussian Field-Marshal-General and War Minister Albrecht Graf von Roon (1803–1879) had already talked about the army as a 'large elementary school', as an 'educational school of the whole nation' in the 1860s. The adoption of this metaphor by Speck also gives a clue to his political attitude. Roon had proposed this idea because he disliked the pre-March-revolutionary concept of a bourgeois oriented militia, *Landwehr*, and thus justified the transformation of the 1814 militia to a royal people's army in 1860. According to Ritter this step expressed a 'clear tendency towards a militarization of the entire people'.[74] Speck went so far as to stylize military orderliness as a quasi-natural property of humans which could be proven in all phases of life:

> The subordination to military order, the ranging into a totality, and the complete merging in this totality is not only an artificial product of outer drill, it is a need deeply rooted in human nature. Who had not seen in our times the very smallest ones who hardly have grown beyond their first exercises in walking, standing in rank and file, and then would have expressed his joy over such an early development of humanely

valuable properties? What games does a growing boy love better than war games? . . . What is it even the simple man prefers to talk about in his later life than his time at the military?[75]

Speck's contribution characterized that common attitude which helped prepare and accompanied World War I. The Imperial politics of the German Reich, which since its unification in 1871 strived for world-political influence, formed the background. Especially after the dismissal of Reich-chancellor Otto von Bismarck in 1890, tensions developed between the European powers of England, France and Russia which ultimately led to World War I.[76]

In parallel there was a worsening foreign policy climate, essentially caused by Germany's highly visible efforts to better equip its army and especially its fleet.[77] This decisively shaped the ideas for education both within and outside schools. They resulted in a disposition for war prepared over many years.[78] Speck's contribution clearly allows us to see how heroes' groves were used for military purposes. Sports activities near heroes' groves served to improve military physical training. The physical presence of the victims was supposed to increase youths' willingness to sacrifice themselves. Nationalist ceremonies in heroes' groves (Figure 11.5) supported the alignment of mass emotion.

The concept of 'youth care', a euphemism, followed purely military purposes. What Pastor had described on a cultural historical level, that is, the creation of a global Germanic empire, had found an equivalent pragmatic level.

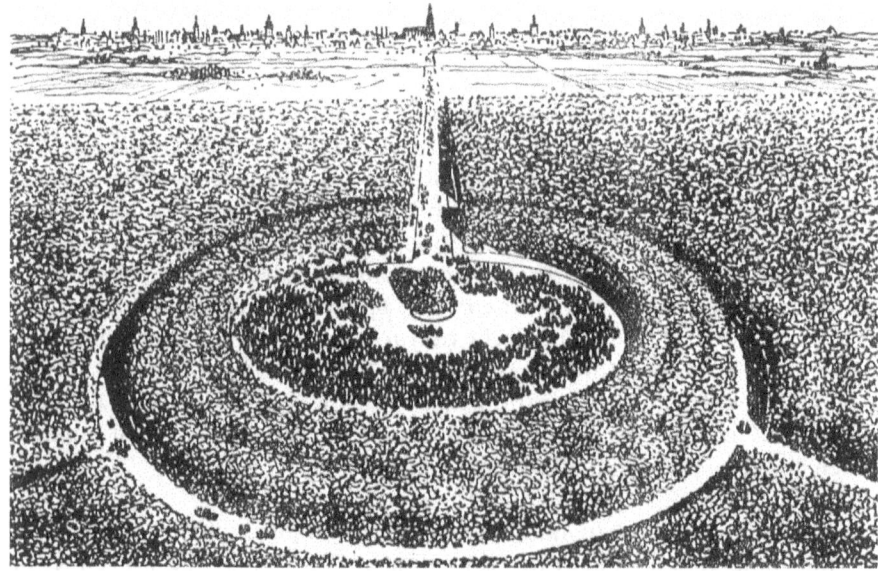

Figure 11.5 Ceremony in a hero grove near a city, drawing Paul Engelhard, after a design by Willy Lange, c. 1916 (Willy Lange, (ed.): 'Arbeitsgemeinschaft für Deutschlands Heldenhaine'. *Deutsche Heldenhaine* (Leipzig, 1915), Nachtrag I, p. 7).

Heroes' groves as holy places, a summary

The dimensions of politics and *Weltanschauung* within the concept of the heroes' groves have been pointed out. German landscape architect Willy Lange connected the traditional symbol of the oak tree to a nationalistic-monarchistic context: the heroes' groves would not only commemorate fallen soldiers but at the same time embodied the emperor and the monarchic principle through the linden planted at their centre. With the shape of the heroes' groves, Lange referred to holy groves in Germanic Antiquity, trying to legitimize continuity in German history from Germanic primeval times which culminated in actual world-political claims of Germany during World War I.

The cultural historian Pastor supported the same principle. The characteristic ring motif for heroes' groves, known from stone-age sun-sanctuaries, is due to him. He considered the ring motif as well as the *Walburg* as derived from religious testimony of an earlier Germanic world empire. Pastor considered its Aryan inhabitants culturally and racially superior to other peoples. The transfer of a characteristic built form from Germanic primeval times legitimized the world-political claims of Germany, both for him and for Lange. The interpretation of history thus served a real political cause.

The legitimatory principle of Lange and Pastor was pragmatically applied to the design proposal by Johannes Speck. Speck utilized the heroes' groves for military purposes. They should raise the readiness of the youth for fighting, both physically and mentally, and were thus immediately useful for Germany's efforts to become a world power. The physical presence of the fallen soldiers glorified as heroes served to guarantee a future readiness for sacrifice. National ceremonies in heroes' groves heightened military purposes politically.

The concept of the heroes' groves was developed according to the political and military interest of Germany in becoming a world power. The original idea behind the concept was that heroes' groves were locations where the fallen could be personally commemorated and remembered by their descendants. This idea had to step back almost entirely behind the ideological manipulation which followed. Instead the warriors symbolized as oaks served as examples for future readiness for sacrifices. The concept proved unusually successful and resonated strongly not only in state authorities but also in individuals. This may have to do with the way it was staged.

The ideological and political purposes of the 'heroes' groves concept' were pseudo-sacredly disguised, making such places 'sacred'. For orientation it is appropriate here to refer to the classical terms Rudolf Otto and Gustav Mensching applied in their treatises on the phenomenology of religion when they considered the sacred, *Das Heilige*.[79]

For Pastor, a hero grove evoked religious veneration similar to that of the holy grove of the Germanic people about which Tacitus had written.[80] In the spirit of the 'mysterium tremendum' the hero grove served as a place of fear and veneration. It was a place where the 'tremble-stimulating secret' was

experienced, fulfilling a central criterion for the sacred.[81] The spatial enclosure of the grove by means of walls and ditches would create and ensure its
separation from the profane surroundings. These spatial boundaries were
preconditions for a numinous[82] reality.[83] Since the hero grove was ascribed
the legal status of inviolability[84] it was declared a taboo area and thus fulfilled a further criterion for sacredness.[85] Lange proclaimed the hero grove
as a kind of national holy locale where ceremonies of church and state could
take place.[86] How unsubstantiated his view was is shown in the interchange-
able cultish function of the hero grove:

> Ceremonies should take place; patriotic ones and peoples fests, church-
> fests and special local fests, Turnfests, marksmen-festivals (*Schützen-
> feste*), and warrior-fests. . . . In spring resurrection may be celebrated
> here with all means of clerical dedication, and in autumn thanksgiving
> may find a joyful conclusion here after a serious ceremony at church.[87]

For Lange it would also have been appropriate to place an altar or a
chapel in the centre of the hero grove. His 'hero groves idea arose from Nordic wood-culture-spirit' as opposed to its 'enemies in church, art and horticulture' who 'talked out of a south-alpine civilisation spirit', as Lange said
in a chapter on design in his 1927 book *Gartenpläne* (Garden Designs).[88]
So, for Lange a hero grove would resemble a church or a temple and thus
fulfil another criterion of sacredness.[89] It may not come as a surprise that
Speck hoped heroes' groves, like ancient sacred sites, would become centre
points and *foci* of national and religious life.[90] Due to his interest in the
ancient 'agon' Speck referred to classical places for worship as examples.
Olympia in Greece comes to mind as a place where cultural rituals were
located near places for sports activities.

What may appear as strange references to highly differing traditional rituals shows an almost desperate need to enshrine heroes' groves in a sacred
aura which served dubious racist and military purposes. Pseudo-sacred
forms were used to disguise missing religious content in a way that seems
pathetic to the modern reader.

Notes

1 For bio- and bibliographical information about Willy Lange see Joachim Wolschke-
  Bulmahn and Gert Gröning, 'Willy Lange 1864–1941, German garden writer,
  garden theorist, and landscape architect', in *Chicago Botanic Garden Encyclopedia of Gardens, History and Design*, ed. by Candice A. Shoemaker (Chicago
  and London: Fitzroy Dearborn, 2001), 3 vols., vol. 2, pp. 757–60.
2 Willy Lange, 'Heldeneichen und Friedenslinden', in Willy Lange, *Deutsche
  Heldenhaine* (Leipzig: Weber, 1915), pp. 77–80 (79); reprint of a contribution to
  Tägliche Rundschau, 8 December 1914.
3 See: Annette Maas, 'Politische Ikonographie im deutsch-französischen Span-
  nungsfeld. Die Kriegerdenkmäler von 1870/71 auf den Schlachtfeldern um
  Metz', in *Der politische Totenkult. Kriegerdenkmäler in der Moderne*, ed. by

Reinhart Koselleck and Michael Jeismann (München: Fink, 1994), pp. 195–222. Lurz had a closer look at war memorial concepts in Germany (see M. Lurz, *Kriegerdenkmäler in Deutschland* (Heidelberg: Esprint, 1985), 6 vols.). He briefly explains their general design principles and ideological foundations but does not refer to Lange and Pastor. Willy Pastor (1867–1933), a culture historian, had joined Lange's *Arbeitsgemeinschaft für deutsche Heldenhaine* (Association for Germany's Heroes Groves) in January 1915. Richard discussed heroes' groves from a landscape architecture perspective in a very short chapter of his doctoral dissertation. See Winfried Richard, *Vom Naturideal zum Kulturideal: Ideologie und Praxis der Gartenkunst im deutschen Kaiserreich*, doctoral dissertation, Technische Universität Berlin, Schriftenreihe des Fachbereichs 14 der Technischen Universität Berlin, vol. 19 (Berlin: Technische Universität,1984). Goebel's contribution to this topic is comparably brief (see Stefan Goebel, *The Great War and Medieval Memory: War, Remembrance and Medievalism in Britain and Germany 1914–1940* (Cambridge: Cambridge University Press, 2007), pp. 76–9).

4 See Jörg Matthies, 'Eichen und Granitfindlinge', in *Stolperstein der Geschichte, Die Ruine des Kieler U-Boot Bunkers als Mahnmal und Herausforderung*, ed. by Jens Rönnau (Kiel: Kilian, 1997), pp. 192–96 (196).

5 Willy Lange,'Die leitenden Gestaltungsgedanken für die Heldenhaine', in Lange, *Deutsche Heldenhaine*, pp. 5–12 (5).

6 See Klaus Lindemann, 'In den deutschen Eichenhainen weht und rauscht der deutsche Gott', in *Der Wald in Mittelalter und Renaissance, Studia Humaniora*, ed. by Josef Semmler (Düsseldorf: Droste, 1991), vol. 17, pp. 200–39; Annemarie Huerlimann,'Die Eiche, heiliger Baum deutscher Nation', in *Waldungen: Die Deutschen und ihr Wald*, Ausstellungskatalog (Berlin: Bernd Weyergraf, 1987), vol. 149, pp. 62–8; Ernst Gall and L. H. Heydenreich (eds.), *Reallexikon zur deutschen Kunstgeschichte* (Stuttgart: Metzler, 1958), vol. 4, columns 914–22; H. Reling and J. Bohnhorst, 'Unsere Pflanzen nach ihrem deutschen Volksnamen, ihre Stellung in Mythologie und Volksglauben, in Sitte und Sage', in *Geschichte und Literatur*, Gotha: E.F. Thienemann,1898; P. Wagler, *Die Eiche in alter und neuer Zeit. Eine mythologisch-culturhistorische Studie* (Berlin, 1891), 2 vols.; Alfred Detering, *Die Bedeutung der Eiche seit der Vorzeit* (Leipzig: Curt Kapitzsch,1939).

7 Eberhard Weis, *Der Durchbruch des Bürgertums, 1776–1847: Propyläen Geschichte Europas* (Frankfurt am Main: Propyläen, 1982), vol. 4, pp. 358–60.

8 See Huerlimann,'Die Eiche, heiliger Baum deutscher Nation', pp. 66–7.

9 Lange, *Deutsche Heldenhaine*, p. 6; see also, p. 80.

10 Huerlimann,'Die Eiche, heiliger Baum deutscher Nation', p. 80.

11 Huerlimann,'Die Eiche, heiliger Baum deutscher Nation', p. 8.

12 Huerlimann,'Die Eiche, heiliger Baum deutscher Nation', pp. 63–72 (67).

13 Huerlimann,'Die Eiche, heiliger Baum deutscher Nation', p. 67.

14 Thomas Nipperdey, "Nationalidee und Nationaldenkmal in Deutschland im 19. Jahrhundert", in Jutta Schuchard and Horst Claussen (eds.), *Vergänglichkeit und Denkmal: Beiträge zur Sepulkralkultur*, (Bonn: Bouvier Verlag, 1985), pp. 189–231.

15 Eric Hobsbawm and Terence Ranger (eds.), *The Invention of Tradition* (Cambridge: Cambridge University Press, 1996), pp. 1–14, 273–79.

16 See R. M. Meyer, *Altgermanische Religionsgeschichte* (Berlin: Quelle & Meyer, 1909), p. 424; Jacob Grimm, *Deutsche Mythologie* (1875–1878) (Wiesbaden; Drei Lilien, reprint 1992), 3 vols., vol. 1, pp. 53–71.

17 Lange, *Deutsche Heldenhaine*, p. 6.

18 Publius Cornelius Tacitus. *Germania. Übersetzung, Erläuterung und Nachwort von Manfred Fuhrmann* (Stuttgart: Reclam, 1995).

19 Lange, *Deutsche Heldenhaine*, p. 9.
20 Willy Lange could have based his hints on Germanic religious beliefs upon a wealth of scholarly publications. See for example the overview given by Meyer, *Altgermanische Religionsgeschichte*, pp. 567–92; 592–632.
21 Ekkehard Mai. 'Die Denkmäler im Kaiserreich', in *Wilhelm Kreis: Architekt zwischen Kaiserreich und Demokratie 1873–1955*, ed. by Winfried Nerdinger and Ekkehard Mai (München, Berlin: Klinkhardt & Biermann, 1994), pp. 28–43.
22 Lange, *Deutsche Heldenhaine*, p. 77.
23 Lange, *Deutsche Heldenhaine*, p. 77.
24 Pastor and Lange seem to have known each other since 1907. In those days Pastor had reported about articles which attacked Lange's activities in an article in the *Tägliche Rundschau* newspaper. Avenarius, the editor of the journal *Kunstwart*, and Arthur Glogau, the officer of the *Deutsche Gesellschaft für Gartenkunst* had turned against Lange in the October and November issues 1907 of *Kunstwart*. See a summary of the event in: An., 'Zeit- und Streitfragen', in *Die Gartenwelt*, 12: 8 (1907), pp. 93–4.
25 Willy Pastor, 'Die Bedeutung des Ringes im Heldenhain', in Lange, *Deutsche Heldenhaine*, pp. 13–15 (15).
26 See Ludger Alscher, Harald Olbrich, et al. (eds.), *Lexikon der Kunst* (Leipzig: Seemann, 1968–1978), 5 vols., vol. 3, p. 243; S. Reden, *Megalith-Kulturen. Zeugnisse einer verschollenen Urkultur* (Köln: DuMont, 1978), pp. 305–11.
27 See a short entry to megalith culture in: *Brockhaus Konversations-Lexikon* (Mannheim: Brockhaus, 1902), vol. 11, p. 720, see 'Megalithische Denkmaeler'. 'Man kennt sie in den meisten Laendern Europas, im noerdlichen Afrika und in Asien'.
28 He opposed other positions which were held in scholarly publications. See: Pastor, 1907: Willy Pastor, *Aus germanischer Vorzeit* (Wittenberg: A. Ziemsen, 1913), p. 175.
29 Willy Pastor, *Altgermanische Monumentalbaukunst*, (Leipzig: Werdandi-Bücherei, 1910), vol. 4, pp. 9–10.
30 Pastor, *Aus germanischer Vorzeit*, p. 175.
31 Walter Jens (ed.), *Kindlers neues Literaturlexikon* (München: Kindler, 1988), 21 vols., vol. 6, p. 406 (406–12), see 'Gobineau'.
32 Jens 1988, vol. 6, p. 406 (406–12), see 'Gobineau'.
33 See Joseph Arthur Comte de Gobineau. *Versuch über die Ungleichheit der Menschenracen*. German edition by Ludwig Schemann. (Stuttgart: Frommann, 1898–1901), 4 vols., vol. 1, pp. 156–95.
34 See de Gobineau 1898, vol. 1, pp. 278–95; 1899, vol. 2, pp. 367–69 and 375; 1901, vol. 4, p. 308.
35 See de Gobineau 1899, vol. 2, pp. 378–82; 1901, vol. 4, p. 309.
36 Houston Stewart Chamberlain: *Die Grundlagen des 20. Jahrhunderts*, (München: Bruckmann, 1899, 1905), 2 vols.: see Jens 1988, vol. 3, pp. 856–57, see 'Chamberlain'.
37 See Pastor, *Altgermanische Monumentalbaukunst*, p. 10.
38 Pastor, *Altgermanische Monumentalbaukunst*, p. 143.
39 Lange, *Deutsche Heldenhaine*, p. 44.
40 See the design by Genzmer and Fader in Lange, *Deutsche Heldenhaine*, pp. 44–5.
41 See also Willy Pastor, *Aus germanischer Vorzeit*, (Wittenberg: A. Ziemsen, 1913), pp. 84–93, here, p. 84; see Lange, *Deutsche Heldenhaine*, pp. 41–2.
42 Lange, *Deutsche Heldenhaine*, pp. 43–4.
43 Pastor 1913, pp. 200–1 and 185.
44 Lange, *Deutsche Heldenhaine*, p. 78.
45 Lange, *Deutsche Heldenhaine*, p. 79.
46 Lange, *Deutsche Heldenhaine*, p. 79.
47 See J. Speck, *Die wissenschaftliche und pädagogische Weiterbildung der akademisch gebildeten Lehrer* (Leipzig, Quelle & Meyer, 1917).

48 Johannes Speck, 'Heldenhaine und Jugendpflege', in Lange, *Deutsche Helden-haine*, pp. 20–31 (20).

49 See Gert Gröning and Joachim Wolschke, 'The youth movement of the Weimar Republic and its understanding of nature', in *Children's Environments Quarterly* 2:2 (1985), pp. 2–6; see also Gert Gröning and Joachim Wolschke, 'Soziale Praxis statt ökologischer Ethik, Zum Gesellschafts- und Naturverständnis in der Jugendbewegung unter besonderer Berücksichtigung der Arbeiterjugendbewegung', in *Archiv der deutschen Jugendbewegung, Jahrbuch* 15/1984–85 (1986), pp. 201–52, Burg Ludwigstein; see also Joachim Wolschke-Bulmahn, 'Auf der Suche nach Arkadien, Zu Landschaftsidealen und Formen der Naturaneignung in der Jugendbewegung und ihrer Bedeutung für die Landespflege', in *Arbeiten zur sozialwissenschaftlich orientierten Freiraumplanung* (München: Minerva, 1990), vol. 11.

50 Speck, 'Heldenhaine und Jugendpflege', in Lange, *Deutsche Heldenhaine*, pp. 26–9.

51 See 'Verhandlungen über Fragen des Höheren Unterrichts, Berlin, 4.-17. Dezember 1890', im Auftrag des Ministers der Geistlichen, Unterrichts- und Medizinal-angelegenheiten, Berlin.

52 Heinz Lemmermann, *Kriegserziehung im Kaiserreich: Studien zur politischen Funktion von Schule und Schulmusik 1890–1918* (Lilienthal/Bremen: Eres, 1984), 2 vols., vol. 1, p. 17.

53 Lemmermann, *Kriegserziehung im Kaiserreich*, p. 18.

54 Lemmermann, *Kriegserziehung im Kaiserreich*, p. 19. A number of contributions to this conference, especially from the military, wanted military strength to be represented more explicitly than earlier on. Lemmermann, *Kriegserziehung im Kaiserreich*, pp. 20–2.

55 Lemmermann, *Kriegserziehung im Kaiserreich*, p. 20.

56 Lemmermann, *Kriegserziehung im Kaiserreich*, p. 21.

57 Ibid., p. 21.

58 Lemmermann, *Kriegserziehung im Kaiserreich*, p. 22.

59 Lemmermann, *Kriegserziehung im Kaiserreich*, p. 30.

60 C. Berg(ed.), *Von der Reichsgründung bis zum Ende des Ersten Weltkriegs: Handbuch der deutschen Bildungsgeschichte* (München: C.H. Beck, 1991), 6 vols., vol. 4 (1870–1918), p. 505.

61 See H. Lorenz, "Wehrkraft und Jugenderziehung: Zeitgemäße Betrachtung auf Grund seines beim Deutschen Kongreß zu Königsberg am 25. Juni 1899 gehaltenen Vortrags (1899)", in *Von der Reichsgründung bis zum Ende des Ersten Weltkriegs*, ed. by Christa Berg (München: C.H. Beck, 1991), p. 505.

62 Lorenz, "Wehrkraft und Jugenderziehung", p. 507.

63 Lemmermann, *Kriegserziehung im Kaiserreich*, p. 31.

64 Lemmermann, *Kriegserziehung im Kaiserreich*, p. 31.

65 Lemmermann, *Kriegserziehung im Kaiserreich*, p. 31.

66 Hermann Lorenz, *Wehrkraft und Jugenderziehung* (Leipzig: Voigtländer, 1899); Hermann Lorenz and Emil Gustav Theodor von Schenkendorf, *Wehrkraft durch Erziehung* (Leipzig: Voigtländer, 1904).

67 Berg, (ed.) *Von der Reichsgründung bis zum Ende des Ersten Weltkriegs*, p. 509.

68 Ibid., p. 510.

69 Ibid., p. 511.

70 See Wolschke-Bulmahn, 'Auf der Suche nach Arkadien', pp. 202–16 (chapter 5.2: Das Kriegsspiel als 'ganzheitlicher' Beitrag vormilitaerischer Ausbildung [war games as 'holistic' contribution to pre-military education]).

71 See his reference to Lorenz and von Schenkendorf 1905; Speck, 'Heldenhaine und Jugendpflege', in Lange, *Deutsche Heldenhaine*, p. 31.

72 See a contemporary definition in the *Handbuch für sozialdemokratische Wähler* (manual for social-democratic voters) from 1898: 'The notion of "militarism" is

not just a catchword as it may appear when one has a superficial look at it. This notion not only comprehends the existing military system as it has gradually developed in Germany but also the spirit, the habits and attitudes of the existing military system which have influenced the entire public and social life, and which in an unfortunate way gains more and more prevalence'; quoted after Jürgen Kuczynski, *Geschichte des Alltags des deutschen Volkes, 1871–1918* (Köln: Papy Rossa, n.d.), 5 vols., vol. 4, p. 314.

73 Speck, 'Heldenhaine und Jugendpflege', in Lange, *Deutsche Heldenhaine*, p. 23.

74 Ritter, G. 1981, 'Von Boyen bis Roon. Volksheer oder königliche Garde', in *Moderne preussische Geschichte 1648–1947*, ed. by O. Büsch and W. Neugebauer. Veröffentlichungen der Historischen Kommission zu Berlin 52:2 (1981), p. 858.

75 Speck, 'Heldenhaine und Jugendpflege', in Lange, *Deutsche Heldenhaine*, pp. 23–4.

76 T. Schieder, *Staatensystem als Vormacht der Welt, 1848–1918*. Propyläen Geschichte Europas (Frankfurt am Main: Propyläen, 1982), 5 vols., pp. 272–92.

77 Schieder, *Staatensystem als Vormacht der Welt, 1848–1918*, pp. 267–69.

78 Berg (ed.), *Von der Reichsgründung bis zum Ende des Ersten Weltkriegs*, pp. 501–14; 515–23 ('Militär und Militarisierung; Der Einfluss des Militärs auf Schule und Lehrerschaft'); Lemmermann, *Kriegserziehung im Kaiserreich*, pp. 11–50.

79 See R. Otto, *Das Heilige: über das Irrationale in der Idee des Göttlichen und sein Verhältnis zum Rationalen* (Breslau: Trewendt & Granier, 1917; 1958), 2 vols; Gustav Mensching, *Die Weltreligionen* (Wiesbaden: VMA, 1996), pp. 283–309.

80 Pastor 1915, p. 13.

81 Mensching, *Die Weltreligionen*, p. 296.

82 Mensching, *Die Weltreligionen*, p. 285; see the definitions for *Numen* and *Numinosum* in *Lexikon der letzten Dinge*, ed. by W. Beltz (Augsburg: Pattloch, 1993), pp. 313–14. The notion of *numen* was defined by Festus, a grammarian of the second century. It signifies the will and power of a god. The *numinosum* signifies an object of human experience which evokes a characteristic reaction which at the same time rests upon the 'attractedness', the fascination, and the 'tremendum', the fear.

83 Mensching, *Die Weltreligionen*, p. 285.

84 Lange, *Deutsche Heldenhaine*, p. 9.

85 Mensching, *Die Weltreligionen*, p. 297.

86 Lange, *Deutsche Heldenhaine*, pp. 6–7.

87 Lange, *Deutsche Heldenhaine*, pp. 6–7.

88 Willy Lange, *Gartenpläne* (Leipzig: Weber, 1927), p. 7.

89 Mensching, *Die Weltreligionen*, p. 309.

90 Speck, 'Heldenhaine und Jugendpflege', in Lange, *Deutsche Heldenhaine*, p. 27.

12 Dan Kiley
Groves, space and complexity

Marc Treib

Perhaps our first associations with the word 'grove' are more mythic and poetic than factual: something involving trees, perhaps with a spring and a nymph or two, or even a baptism in a painting of the Italian Renaissance. To those of us who have lived at least some fragments of our lives in subtropical Florida (as I have), the word grove instead triggers associations with oranges and orange trees, long ranges of them that blanket in all directions the slightest of slopes, with trees set in regular intervals that induce axial as well as diagonal views. Regularity also appears in poplar groves grown for commercial purposes in France, and the *pinetum* that has long been a part of the historic Italian garden.[1] Or perhaps the grove serves even as a *memento mori* in a familial cemetery in Tuscany (Figure 12.1). These two manners, the natural and the geometric, stake out the territory of the grove,

Figure 12.1 Cypress grove, Tuscany, Italy (Marc Treib).

and both have continuously appeared throughout at least two millennia of landscape architecture. Most of the chapters in this book address the forest and the tree. Instead, I would like to look at trees in groves designed in the twentieth century, focusing on the *spaces* that have resulted from their planting, mostly in a regular order. Here the work of Dan Kiley is demonstrative.

Daniel Urban Kiley (1912–2004) was born in south Boston in 1912 and matured, studied, and worked in New England.[2] Kiley was a true Yankee, and truly Irish: practical more than theoretical, a devoted craftsman, rigorous in his designs and in their details, highly opinionated, and a consumer of considerable amounts of gin. His early training veered toward architecture, but an apprenticeship in the Boston office of the landscape architect Warren Manning (1860–1938) changed the course of his studies, and ultimately his life's work. In terms of planning and planting, Manning's designs were accomplished, but in terms of formal or spatial innovation they were rarely distinguished.[3] He worked in the manner of the time, which was eclectic, bending with the conditions of client and terrain. In all, Manning's landscapes were more about propriety than invention and more about species than spaces. Despite Manning's urgings to the contrary, Kiley enrolled in the graduate landscape architecture programme at Harvard's Graduate School of Design, not in its degree programme but as a special student. He endured formal education only a short time, about two years, but during that time he fell in with James Rose and Garrett Eckbo, two other young and restless landscape graduate students. The trio is often credited for having brought modernism to American landscape architecture, but, of course, the story is not really that simple, as a number of other landscape architects were also pursuing the same goals.

The grove and the bosk played no real role in Kiley's own pre-World War II professional landscape designs, which often took form as a collage of modernistic motifs. But European military service – he was the principal designer of the main courtroom for the war trials in Nuremburg – brought him into contact with garden cultures new to him, garden cultures that included the work of André le Nôtre in France. That encounter was to prove consequential. In Le Nôtre's work, particularly the gardens of Vaux le Vicomte and Sceaux, Kiley discovered the beauty of order and its ability to structure, and humanize, space. In the allée and the bosk he found the means to configure the landscape and modulate its scale. At an increasing rate his postwar works spurned the feigned naturalism of the Olmsted tradition which then dominated, and still dominates, American landscape architecture, at least in projects of a certain scale, and the more complex geometries used by contemporaries like Garrett Eckbo. Today, Dan Kiley might be called a 'modern classicist' or 'classical modernist'; I prefer the latter.

Like Le Corbusier, for example in his *Poem of the Right Angle*, Kiley did not position geometry in opposition to nature.[4] Being a product of the human mind, geometry was therefore, by definition, natural. Humans were a part of nature. Like the painter Jackson Pollock, who probably did say it, Kiley could have said 'I am nature'.[5] In fact Kiley once did say, 'It's not man *and* nature. It's not man *with* nature. Man *is* nature, like the trees'.[6] As early as 1913 the German landscape architect Leberecht Migge had hailed geometry as the means by which to escape the lure of naturalistic planting. In *Garden Culture in the Twentieth Century*, Migge wrote: 'Nature is never tasteless. But it is not always interesting'.[7] Instead, he asserted that 'the regularity and control of the geometric line promises a stronger, more rhythmic effect in opposition to the arbitrariness of the free'.[8] Kiley would have agreed. Henry Arnold had worked for Kiley for 25 years before he wrote *Trees in Urban Design*, first published in 1982. Obviously influenced by his time in the Kiley office, Arnold stressed that 'human artistry can improve on rural nature by shaping the materials of the city, including trees, to create a better urban habitat than now exists, without copying "nature"'.[9] A selection of landscapes illustrates Kiley's masterful use of the geometrically planned grove and demonstrates how, over time, the spaces within these groves became more complex and more engaging.

The productive orchard has been an element of the New England landscape since its settlement in the seventeenth century, brought to the New World by Puritan and other European settlers. Kiley frequently employed this landscape type in the postwar years and throughout his later years of practice. At the 1959 Currier Farm in Danby, Vermont, a low wall bounded the apple orchard, used as a vital part of the renovated landscape. The natural fall of the terrain required setting the garden's various zones on several levels (Figure 12.2). To a degree greater than in other works both before and later, the various garden zones of the Currier Farm were treated independently (elements of the garden, such as the orchard, pre-existed Kiley's design and were incorporated into the new plan). As a whole the project was less than the integrated whole characteristic of projects from just a few years before, or in projects to come.

Some six years later, in a small town in Indiana, Kiley created the Hamilton garden. By that time, 1965, both the landscape and buildings of Columbus, Indiana, had been consequently and positively affected by the establishment of a private foundation that paid the design fees for new public work, although only if the city's building committee selected its architect from a list provided by the foundation.[10] This gracious gesture brought an impressive array of architectural talent to Columbus, for example Eliel Saarinen in the late 1940s, his son Eero in the 1960s, IM Pei, Skidmore, Owings and Merrill, and a host of others over a period of now half a century.[11] As a result, this small town of 43,000 is today a museum of modern architecture and a pilgrimage destination for those interested in design. Dan Kiley became the consulting landscape architect for the town and designed many of its landscapes, both private and public.

The Hamilton garden, although a private commission, stemmed from this professional connection. A house already occupied the site, as did a garden. In the revised design Kiley used the allée to structure the zones of the

Figure 12.2 Dan Kiley, Currier Farm, Danby, Vermont, 1959. Orchard, stone walls and paving (Marc Treib).

Figure 12.3 Dan Kiley, Hamilton garden, Columbus, Indiana, 1965. View from the pavilion toward the poolhouse (Marc Treib).

small garden. With the pool and terrace set to the north as an extension of the house, the perpendicular allée paired pergolas set at the opposite edges of the lot (Figure 12.3). The little leaf lindens (*Tilia cordata*) were planted directly in lawn, with no other suggestion of a path on the ground plane than the allée itself. What distinguishes a Kiley bosk or grove from those of, say, Le Nôtre, however, is its transparency. Unlike the vegetal masses of the *bosquets* at Versailles, where blocks of shrubbery structure and confine the paths, Kiley almost always pruned his trees sufficiently high to see beneath them and to allow physical passage. This treatment would remain nearly ubiquitous in his designs.

The 1988 Henry Moore Sculpture Garden spread outward and downward from the mass of the classical Nelson Atkins Art Museum in Kansas City, Missouri. Five grassed terraces, which Kiley called 'shelves', each edged with Japanese yew, modulated the sloping land as it falls to the river.[12] A grove of ginkgo trees punctuated those terraces intended for the display of smaller sculptures (Figure 12.4). To bind the great central space and screen adjacent buildings and neighbours, Kiley planted allées of Redmond linden (*Tilia americana* 'Redmond') on either side of the lawn. To increase the visual density provided by a typical allée or square of four trees, Kiley added a fifth tree at its centre, thereby creating a quincunx. This ordering, a long-known landscape design device, provided greater visual opacity when

Figure 12.4 Dan Kiley, Henry Moore Sculpture Garden, Nelson Atkins Art Museum, Kansas City, Missouri, 1989. Terraces outlined with hedges and the grove of irregularly spaced ginkgos (Marc Treib).

viewed from a perpendicular position and diagonal views from others, especially when seen in movement. Greater sensual complexity also resulted. At the base of the trees, shrubbery created a transition layer between the ground plane and the lower limit of the tree canopy.

Kiley's ability to modulate complexity had increased noticeably in his garden for Irwin and Xenia Miller, also in Columbus, a garden to accompany their modernist house designed by Eero Saarinen and first occupied in 1957. Here the planning and use of the grove became more involved and the spatial effects more varied, with the readings changing with viewpoint and promenade, and with the seasons of the year. The plan of the house relied on a pinwheel arrangement of four functional areas that pivot around the central living room. Smaller gardens complement the rooms that extended outwards from the central area toward the sunlight. To the north a grove of multi-trunk redbuds (*Cercis canadensis*), which were later changed to crape myrtles (*Lagerstroemia*), provided those using the dining room with visual interest and a gauge by which to trace the flow of seasons. Critical to the linkage of house and garden was a ten-foot wide podium that extended the white terrazzo floor of the interior outward, ultimately transformed into ground cover or lawn within the garden. The justifiably celebrated allée of honey locusts (*Gleditsia triacanthos*) was once terminated at one end by a now-departed recumbent figure by Henry Moore (Figure 12.5).[13] Although the gravel-paved allée supports afternoon promenades along the edge of the bluff, it functions more significantly when viewed perpendicularly, as a screen or filter between the intensively maintained garden and the lawn and woods on the floodplain below. In effect, the honey locusts act as a secondary layer between the living room and the landscape beyond, a living screen that confronts and softens the afternoon sun in the summer months, while in winter enhancing the view to the river in the distance.

A wall of arborvitae screens the Miller compound from public streets on two sides of the property. Rather than a single uniform hedge, Kiley cut the green wall into segments, each unit battered for increased exposure to sunlight and staggered as a rhythmic sequence that undermines its apparent length. The entrance to the grounds originally followed an allée of horse chestnut trees (*Aesculus hippocastanum*); these have recently been replaced with a species of Ohio buckeye (*Aesculus glabra*). East of the house, Kiley developed the grounds by counterpoising the volumes of two apple orchards, while lacing them together with a row of red maples (*Acer rubrum*). Closer to the house, a second alignment, this time of taller white oaks (*Quercus alba*), distinguishes the respective zones for architecture and garden (Figure 12.6).

The groves of the Miller garden apply two tropes characteristic of modernist landscape design. In the first, tree rows of differing interspacing and heights are treated as planes and masses rather than as lines and are used to articulate spaces of varying density and visual opacity. The second characteristic I have termed 'slippage' in an earlier essay.[14] By this I mean that

Figure 12.5 Dan Kiley, Miller garden, Columbus, Indiana, 1957. Honey locust allée (Marc Treib).

Figure 12.6 Dan Kiley, Miller garden, Columbus, Indiana, 1957. Layers, volumes and spaces of the apple orchard and row of white oaks (Marc Treib).

in most instances, the alignments of groves and voids across a site inter-
penetrate rather than abut, and that their potential meetings at the corner
are left open. This practice allows the space to 'leak' or 'slip' from one
spatial unit to the next, from the outside in or the inside out. Mies van der
Rohe provided the architectural model for this treatment over three decades before in his German Pavilion for the 1929 international exposition
in Barcelona. Within the building and its walled courtyards veneered with
exotic stones, the dynamic arrangement of free-standing planes composed a
gracious spatial fluidity. Kiley adopted this idea of interpenetrating spaces, a
hallmark of the modern manner, but executed it in vegetal materials. Where
Mies was restricted by masonry walls and a constant ceiling height, Kiley
could profit from the differences in volume and enclosure effected by colum-
nar tree trunks, dense hedges, and multi-stemmed shrubs, and, of course,
seasonal change. Garrett Eckbo had proposed such uses of trees as early as
the late 1930s in his designs for a series of small parks at migrant agricultural workers' camps in California's Central Valley.[15] Using these methods
Kiley fashioned complex, multivalent spaces using formal, often classical,
elements, but spaces and vegetation firmly rooted in the twentieth century.

In many ways, Kiley's several-decades long investigation of the grove culminated in the NCNB (now Nationsbank) Plaza in Tampa, Florida, completed in 1988. Perhaps calling the space a 'plaza' is misleading, as in many
respects it is more a garden than what we normally consider a plaza. In
addition, the four-and-a-half acre wedge of landscape covers the roof of a
parking garage and stands a full floor above street level removed from direct
access. The 33-story office tower and cubic bank facility by the architect
Harry Wolf were planned using dimensions based on the Fibonacci progression of numbers, where the sum of the two prior figures yields the next in
the series [Figure 12.7]. Kiley continued that proportional system outdoors
and used that mathematical progression to determine the interweaving of
tough zoyzia grass with slabs of concrete paving.

The plaza, or garden, or simply terrace (whatever we prefer to term it)
interlaced water, paving, ground cover, flowering shrubs, and trees, using
their heights and their specific formal properties to weave an incredibly rich
spatial and patterned tapestry. At street level, two reflective sheets of water
foregrounded the street-side walls of the parking garage. The entrance cor-
ridor to the garage was roofed in glass and served as the landscape's primary
irrigation canal, which was set nearly parallel to the Hillsborough River
which bounded the site on its opposite side. From this source, narrow rills
penetrated the field of grass and paving, and terminated as round basins
with bubblers recalling those at the Court of the Myrtles in the Alhambra.
That tiny historicist touch was the garden's sole reference, although the rep-
etition of rills suggests, at least in a distant way, the Patio de los Naranjos in
either Seville or Cordova.

As noted before, the proportion of growing surface to paving varies with
the numerical progressions with more green panels as it moves away from

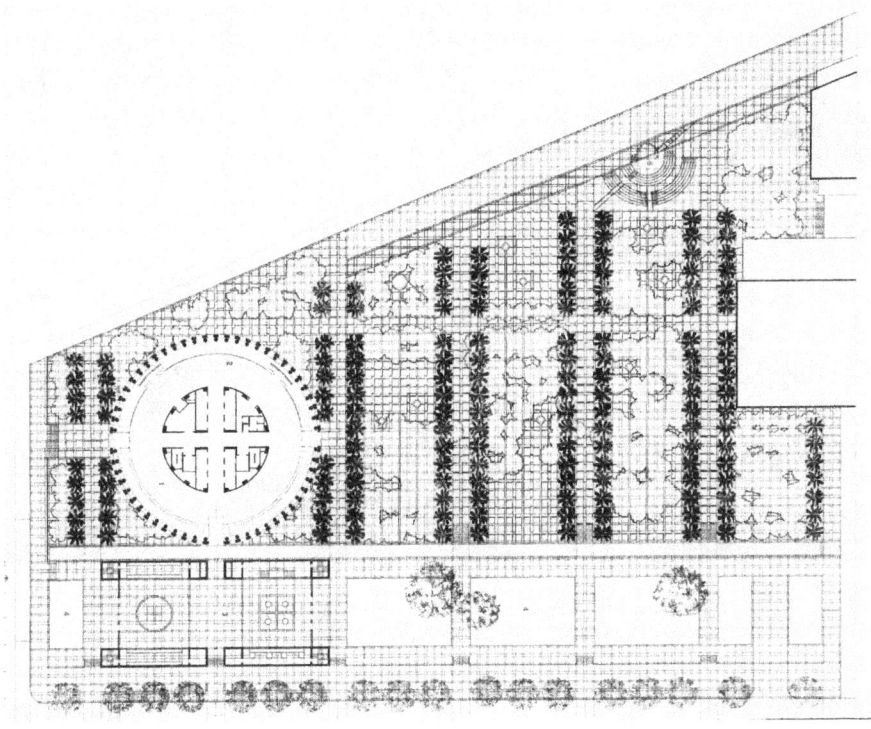

Figure 12.7 Dan Kiley, NCNB Plaza, Tampa, Florida, 1988. Plan (courtesy: Office of Dan Kiley).

the buildings. Upon that 'bed of precision' as Kiley called it, alignments of sabal palms (*Sabal palmetto*) extruded the numerical progression vertically, continuing the garden's ordering structure with trunks that read as columns.[16] Tall in relation to the diameters of their trunks, the palms were an appropriate choice, especially because they bend rather than break in high wind and have root balls that remain quite small, a benefit for a landscape built on structure.

The most brilliant design decision was to create a middle storey to this composition, with leafy volumes set between the high crowns of the palms and the grassed or paved ground plane (Figure 12.8). Still using the same proportional grid, Kiley planted crape myrtles (*Lagerstroemia*) in what appeared to be a free pattern. But of course it was not: it, too, shared the common orthogonal and proportional ordering of the entire site. The resulting effect profited from the play between the lower, dense, more natural-looking plantations and those above, those taller and more regular. In fact, both groves were plotted using the same mathematical thinking. We seek order in these meandering volumes of leaves and temporal colour as we

Figure 12.8 Dan Kiley, NCNB Plaza, Tampa, Florida, 1988. Interweaving of paving, grass and crape myrtle trees (Marc Treib).

order the stars in the night skies into figures we call constellations. The ordering of the palm trees was geometric; the impression that resulted was not.

The NCNB Plaza demonstrates the richness possible for landscape architecture when using the grove as the project's principal design element. The order of its trees may be regular or irregular; its trees and shrubs may be tall or low. They may read as a field of poles capped by a leafy canopy or as a tangled mass of trunks. They may be bounding and opaque, or visually open, modulating space rather than secluding it. In my estimation, the NCNB Plaza was one of the greatest landscapes of the twentieth century and among Kiley's finest works. Left untended and unrepaired it has withered. Despite its passing, however, the Tampa terrace, like so many other Kiley landscapes, shows us the stunning brilliance that may be achieved by mixing groves, water, grasses, and paving, provided, that is, they are configured and interwoven by a talented designer working with a supportive client, given proper maintenance and gaining the appreciation of those who come.

Notes

1 The *pinetum*, a plantation or grove of pine trees or related conifers, dates back to Roman times and was usually planted as a regular grid.
2 The most complete study of Kiley's work, in terms of breadth if not depth, is Dan Kiley and Jane Amidon, *Dan Kiley: The Complete Works of America's Master Landscape Architect* (Boston: Bulfinch, 1999).

3 Among Manning's best-known works is Stan Hywet located outside Akron, Ohio, built in 1915. Its landscape design mixes extensive areas of naturalistic planting in the English manner with formal gardens in closer proximity to the house.

4 Le Corbusier, *Le Poeme de l'Angle Droit* (Poem of the Right Angle) comprises a series of drawings and poems produced between 1947 and 1953.

5 The statement was quoted by Lee Krasner (1964) in her 'Oral History Interview with Dorothy Seckler for the Smithsonian Institution Archives of American Art,' 1964.

6 Dan Kiley, 'Lecture,' in *Dan Kiley Landscapes: The Poetry of Space*, ed. by Reuben Rainey and Marc Treib (Richmond, CA: William Stout, 2009), p. 23.

7 Leberecht Migge, *Garden Culture in the Twentieth Century* (Die Gartenkultur des 20. Jahrhunderts, 1913), trans. David Haney (Washington, DC: Dumbarton Oaks, 2013), p. 139.

8 Migge, *Garden Culture in the Twentieth Century*, p. 71.

9 Henry Arnold, *Trees in Urban Design* (New York: Van Nostrand Reinhold, 1982), p. vii.

10 The Cummins Foundation was founded in 1954, its Architecture Program in 1960. To date it has funded more than 50 projects.

11 For an overview of the project and its architecture see Steven R. Risting, ed., *Columbus Indiana: A Look at Modern Architecture and Art* (Columbus, IN: Columbus Indiana Visitor Center, 2012).

12 Kiley and Amidon, *Dan Kiley: The Complete Works*, p. 112.

13 This is, in fact, the most famous single image of the garden. The Miller art collection was sold at the time the house and grounds were donated to the Indianapolis Museum of Art.

14 Marc Treib, 'Dan Kiley and Classical Modernism: Mies in Leaf,' *Landscape Journal*, 24:1 (2005), pp. 1–12.

15 See Garrett Eckbo, *Landscape for Living* (New York: Duell, Sloan and Pearce, 1950), figures 143, 171–72, 194–224.

16 Eckbo, *Landscape for Living*, p. 109.

Index